HOW THINGS WORK

STRUCTURES

TIME LIFE® BOOKS®

Other publications:

COVER

The Köhlbrand Bridge, which spans Germany's Elbe River near Hamburg, is a beautiful example of the union of concrete and steel in modern structures. Completed in 1975, the entire bridge is more than two miles long; its main span, supported by radiating steel cables from overhead towers and powerful concrete piers, extends 1,066 feet.

HOW THINGS WORK

STRUCTURES

TIME-LIFE BOOKS

ALEXANDRIA, VIRGINIA

Library of Congress Cataloging-in-Publication Data

Structures
        p.   cm. – (How things work)
    Includes index.
    ISBN 0-8094-7898-6 (trade)
    ISBN 0-8094-7899-4 (lib.)
    1. Structural engineering. 2. Civil Engineering.
    I. Time-Life Books. II. Series.
    TA633.S89   1991
    624—dc20   91-2829
                                                CIP

How Things Work was produced by
ST. REMY PRESS

PRESIDENT	Pierre Léveillé
PUBLISHER	Kenneth Winchester

Staff for *STRUCTURES*

Editor	Hugh Wilson
Art Director	Odette Sévigny
Assistant Editor	Heather L. Mills
Contributing Editor	George Daniels
Research Editor	Fiona Gilsenan
Researcher	Hayes Jackson
Picture Editor	Chris Jackson
Designer	Luc Germain
Illustrators	Chantal Bilodeau, Maryse Doray, Nicolas Moumouris, Robert Paquet, Maryo Proulx
Index	Christine M. Jacobs

Staff for *HOW THINGS WORK*

Series Editor	Carolyn Jackson
Senior Art Director	Diane Denoncourt
Senior Editor	Elizabeth Cameron
Researcher	Nyla Ahmad
Administrator	Natalie Watanabe
Production Manager	Michelle Turbide
Coordinator	Dominique Gagné
Systems Coordinator	Jean-Luc Roy

Time-Life Books Inc. is a wholly owned subsidiary of
THE TIME INC. BOOK COMPANY

President and Chief	Kelso F. Sutton
President, Time Inc. Books Direct	Christopher T. Linen

TIME-LIFE BOOKS INC.

Managing Editor	Thomas H. Flaherty
Director of Editorial Resources	Elise D. Ritter-Clough
Director of Photography and Research	John Conrad Weiser
Editorial Board	Dale Brown, Roberta Conlan, Laura Foreman, Lee Hassig, Jim Hicks, Blaine Marshall, Rita Thievon Mullin, Henry Woodhead
PUBLISHER	Joseph J. Ward
Associate Publisher	Trevor Lunn
Editorial Director	Donia Steele
Marketing Director	Regina Hall
Director of Design	Louis Klein
Supervisor of Quality Control	James King

Editorial Operations

Production	Celia Beattie
Library	Louise D. Forstall
Correspondents	Elisabeth Kraemer-Singh (Bonn); Christina Lieberman (New York); Maria Vincenza Aloisi (Paris); Ann Natanson (Rome).

THE WRITERS

Sarah Brash's work has appeared in numerous Time-Life Books series, including the *Life Science Library*, *Voyage Through the Universe*, *Planet Earth* and *Understanding Computers*.

Matthew Cope is a Montreal-based freelance writer and broadcaster covering arts and sciences. He edited *Computers* for the *How Things Work* series.

Charles Foran is a Montreal writer who has contributed to various Canadian magazines and newspapers.

Dónal Kevin Gordon is a Vermont-based freelance writer. He has written previously for Time-Life Books, working on *Mysteries of the Unknown, Understanding Computers* and *The Third Reich*.

Peter Pocock is a freelance writer with a special interest in science and technology. He was a writer and editor for Time-Life Books for 10 years.

THE CONSULTANTS

Dr. Tom F. Peters is an architect and historian of technology at Lehigh University where he is director of the Institute for the Study of the Highrise Habitat. He is an educator in architectural technology and a writer.

Richard Kramer, P.E., is a geotechnical engineer who has worked for the U.S. Bureau of Land Reclamation since 1960, specializing in the design and analysis of embankment dams. He also teaches, writes and lectures on dam design and construction.

Stefan Muszynski earned a Masters degree in experimental nuclear physics from McGill University. He teaches physics at Dawson College in Montreal, Canada, specializing in statics and the strength of materials.

Michael Bailey, P.E., is an associate at Hart Crowser, a geotechnical and environmental consulting company in Seattle, WA. He has worked on tunneling projects such as the Eklunta Water Tunnel in Alaska.

Norman Nadel, P.E., is the chairman of Nadel Associates, Inc., in Brewster, New York, which provides consulting expertise to the heavy construction industry, particularly for underground work.

Dr. Robert Mark is a professor of architecture and civil engineering at Princeton University. He is the author of *Experiments in Gothic Structure* and *Light, Wind and Structure,* among other publications.

For information about any Time-Life book, please write:
Reader Information
Time-Life Customer Service
P.O. Box C-32068
Richmond, Virginia
23261-2068

CONTENTS

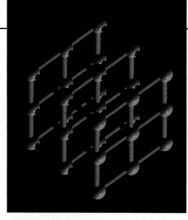

Model of a structural material.

The Sunshine Skyway at Tampa Bay in Florida.

A massive tunnel boring machine.

Spaceship Earth at Walt Disney World in Florida.

The Original Engineers

Through trial and error—and often reason and inspiration—humans learned to plan and build structures. Though the name of the world's first engineer is lost in the mists of prehistory, he and those who came after him gradually assembled a wide body of knowledge—and a stern code of ethics. Hammurabi's code, from 2200 B.C., at least testifies to the penalty for unsound building practices: "If a builder builds a house that is a not firmly constructed and it collapses and causes the death of its owner, that builder shall be put to death."

By 2000 B.C. tremendous strides had been made by engineers in Egypt, Mesopotamia, China and India. They produced sophisticated structures, though little is known about the theoretical foundation for their work; in fact, it remains a mystery exactly how the pyramids were built. Yet it was the Romans who reigned as history's greatest engineers. At their zenith they erected huge arenas, bridges, aquaducts, dams and temples. They did not invent concrete, the arch nor the hammer and chisel. But they understood better than those before them—and for a long time after—how to use the correct materials, tools and techniques to produce lasting monuments to the engineer's vision.

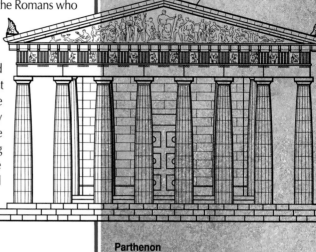

Parthenon
Athens, Greece (432 B.C.)
This temple, parts of which are still standing, was built to honor Athena, the Greek goddess of war and wisdom. Constructed of marble, it measures 101 feet by 228 feet and rests on a limestone base.

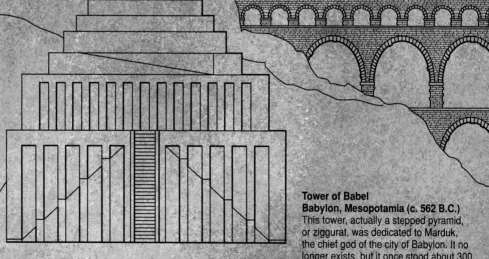

Valley of the Kings
Thebes, Egypt
(c. 1539 - 1075 B.C.)
To prevent robbers from finding the treasure-laden tombs of their pharaohs, the early Egyptians dug their tombs into the natural rock of the mountain overlooking the Valley of the Kings. This necropolis features dozens of tunnels, going as deep as 320 feet, each ending in a burial chamber.

Tower of Babel
Babylon, Mesopotamia (c. 562 B.C.)
This tower, actually a stepped pyramid, or ziggurat, was dedicated to Marduk, the chief god of the city of Babylon. It no longer exists, but it once stood about 300 feet high on a 1.3-acre base and consisted of a 49-foot-thick shell of bricks.

The Great Pyramid
Giza, Egypt (c. 2600 B.C.)
Many massive pyramids were built by the ancient Egyptians as tombs for their pharaohs. The Great Pyramid of Cheops, originally 480 feet high and on a 755-foot-wide square base, was built of more than 2 million limestone blocks, the average block weighing about 2 1/2 tons. The original structure was covered with a smooth white limestone face.

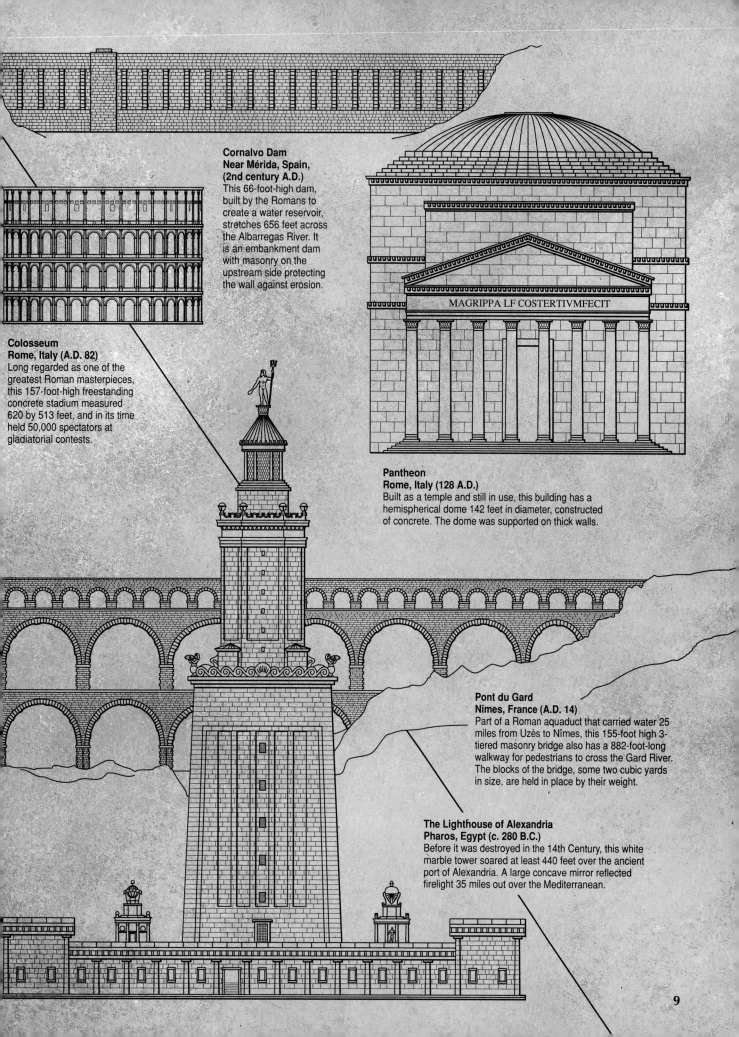

**Cornalvo Dam
Near Mérida, Spain,
(2nd century A.D.)**
This 66-foot-high dam,
built by the Romans to
create a water reservoir,
stretches 656 feet across
the Albarregas River. It
is an embankment dam
with masonry on the
upstream side protecting
the wall against erosion.

**Colosseum
Rome, Italy (A.D. 82)**
Long regarded as one of the
greatest Roman masterpieces,
this 157-foot-high freestanding
concrete stadium measured
620 by 513 feet, and in its time
held 50,000 spectators at
gladiatorial contests.

MAGRIPPA LF COSTERTIVMFECIT

**Pantheon
Rome, Italy (128 A.D.)**
Built as a temple and still in use, this building has a
hemispherical dome 142 feet in diameter, constructed
of concrete. The dome was supported on thick walls.

**Pont du Gard
Nîmes, France (A.D. 14)**
Part of a Roman aquaduct that carried water 25
miles from Uzès to Nîmes, this 155-foot high 3-
tiered masonry bridge also has a 882-foot-long
walkway for pedestrians to cross the Gard River.
The blocks of the bridge, some two cubic yards
in size, are held in place by their weight.

**The Lighthouse of Alexandria
Pharos, Egypt (c. 280 B.C.)**
Before it was destroyed in the 14th Century, this white
marble tower soared at least 440 feet over the ancient
port of Alexandria. A large concave mirror reflected
firelight 35 miles out over the Mediterranean.

9

The Age of Empiricism

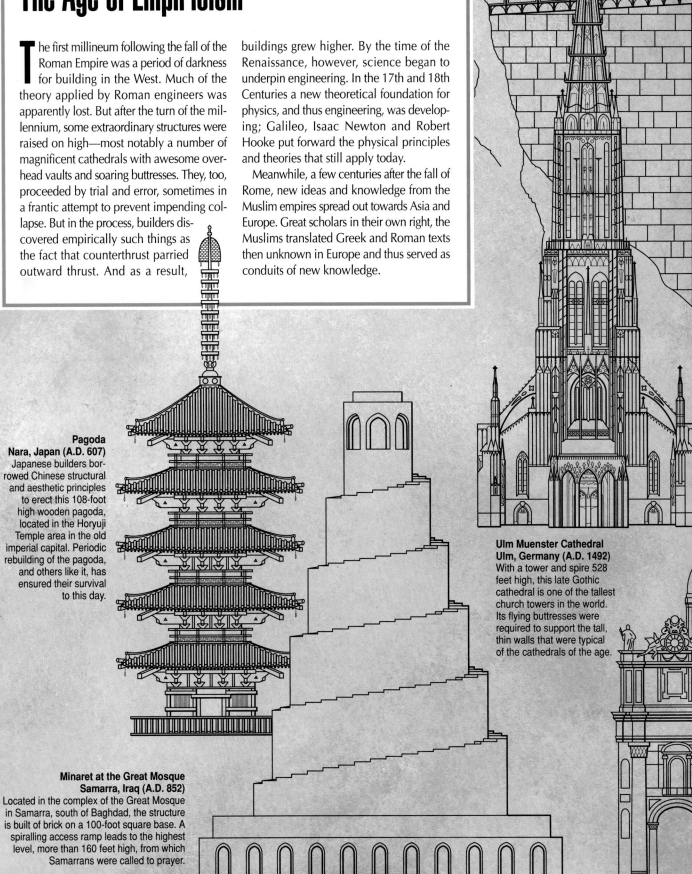

The first millineum following the fall of the Roman Empire was a period of darkness for building in the West. Much of the theory applied by Roman engineers was apparently lost. But after the turn of the millennium, some extraordinary structures were raised on high—most notably a number of magnificent cathedrals with awesome overhead vaults and soaring buttresses. They, too, proceeded by trial and error, sometimes in a frantic attempt to prevent impending collapse. But in the process, builders discovered empirically such things as the fact that counterthrust parried outward thrust. And as a result,

buildings grew higher. By the time of the Renaissance, however, science began to underpin engineering. In the 17th and 18th Centuries a new theoretical foundation for physics, and thus engineering, was developing; Galileo, Isaac Newton and Robert Hooke put forward the physical principles and theories that still apply today.

Meanwhile, a few centuries after the fall of Rome, new ideas and knowledge from the Muslim empires spread out towards Asia and Europe. Great scholars in their own right, the Muslims translated Greek and Roman texts then unknown in Europe and thus served as conduits of new knowledge.

**Pagoda
Nara, Japan (A.D. 607)**
Japanese builders borrowed Chinese structural and aesthetic principles to erect this 108-foot high wooden pagoda, located in the Horyuji Temple area in the old imperial capital. Periodic rebuilding of the pagoda, and others like it, has ensured their survival to this day.

**Ulm Muenster Cathedral
Ulm, Germany (A.D. 1492)**
With a tower and spire 528 feet high, this late Gothic cathedral is one of the tallest church towers in the world. Its flying buttresses were required to support the tall, thin walls that were typical of the cathedrals of the age.

**Minaret at the Great Mosque
Samarra, Iraq (A.D. 852)**
Located in the complex of the Great Mosque in Samarra, south of Baghdad, the structure is built of brick on a 100-foot square base. A spiralling access ramp leads to the highest level, more than 160 feet high, from which Samarrans were called to prayer.

Iron Bridge
Coalbrookdale, England (A.D. 1779)
This 100-foot long bridge across the Severn River was the first iron arch bridge. Its joints are held together without bolts or rivets; the connections, based on carpentry techniques, are slotted and dovetailed.

Tibi Dam,
Tibi, Spain (A.D. 1589)
This 138-foot high masonry dam across the Monegre River was the highest in Europe for almost three centuries. An arch-gravity structure, it worked on the basis of its own weight and the inherent strength of the arch.

Box Tunnel
Box, England (A.D. 1841)
This 1.8-mile long brick-lined tunnel was built between Box and Corsham on the line of the Great Western Railway between Bristol and London. All told 247,000 cubic yards of rock and earth was removed in buckets that were hauled up the access shafts by horses.

PRINCIPIS APOST   PAVLVS V BVRGHESIVS ROMANVS   PONT MAX AN MD CXII PONT VII

St. Peter's Cathedral
Rome, Italy (A.D. 1626)
The main section of this great edifice encloses an area of seven acres. The massive 137-foot-diameter dome was designed by Michelangelo and is so large that the Capitol Building in Washington, D.C. could fit inside with room to spare.

The New Material World

By the 19th Century science and engineering had become inextricably intertwined, resulting in many remarkable advances. But it was the revolution in materials that truly changed the nature of structures. In 1854, Britain's Sir Henry Bessemer discovered how to produce high-quality steel from pig iron by oxidizing out the impurities in great furnaces, and afterwards, Bessemer steel became the foundation of structures that were stronger, more durable and more reliable than ever before.

The development of Portland cement in the early 1800s heralded another material gain; now the properties of mixing and setting concrete could be predicted and controlled. Moreover, the new concretes even could be poured and would set under water. Late in the 19th Century concrete and steel were combined to make reinforced concrete, a marvelously versatile material that took advantage of the strength of both substances. Later, plastics, fiberglass and an assortment of improved metals gave architects and engineers the wherewithal to create even more interesting structures.

But with today's increased scientific knowledge, ongoing materials research and advanced tools and machinery, modern structures span farther and reach higher than most people could have imagined.

Gateway Arch, St. Louis, Missouri (A.D. 1965)
Overlooking the Mississippi River, this monumental 630-foot-high arch is a stressed-skin structure with concrete sandwiched between outer and inner walls of steel. A tramway in the arch's hollow center carries passengers to an observation gallery located at the top.

St. Gotthard Tunnel Switzerland (A.D. 1880)
Using dynamite to blast away solid rock, engineers drove through nine miles of mountain to complete a railway connecting Goeschenen to Airolo. A parallel automobile tunnel was completed in 1980.

Eiffel Tower Paris, France (A.D. 1889)
Built to demonstrate the possibilities of metal in modern construction, this 985-foot-tall tower was erected using then-current knowledge of behavior of materials under loads. Its skeleton is composed of 18,038 iron structural members.

Brooklyn Bridge New York City, New York (A.D. 1883)
Joining Brooklyn to Manhattan across the East River, this famed American suspension bridge spans 1,595 feet between its masonry towers. Bridge builder John Roebling used steel wire instead of iron links for each of the bridge's four main cables.

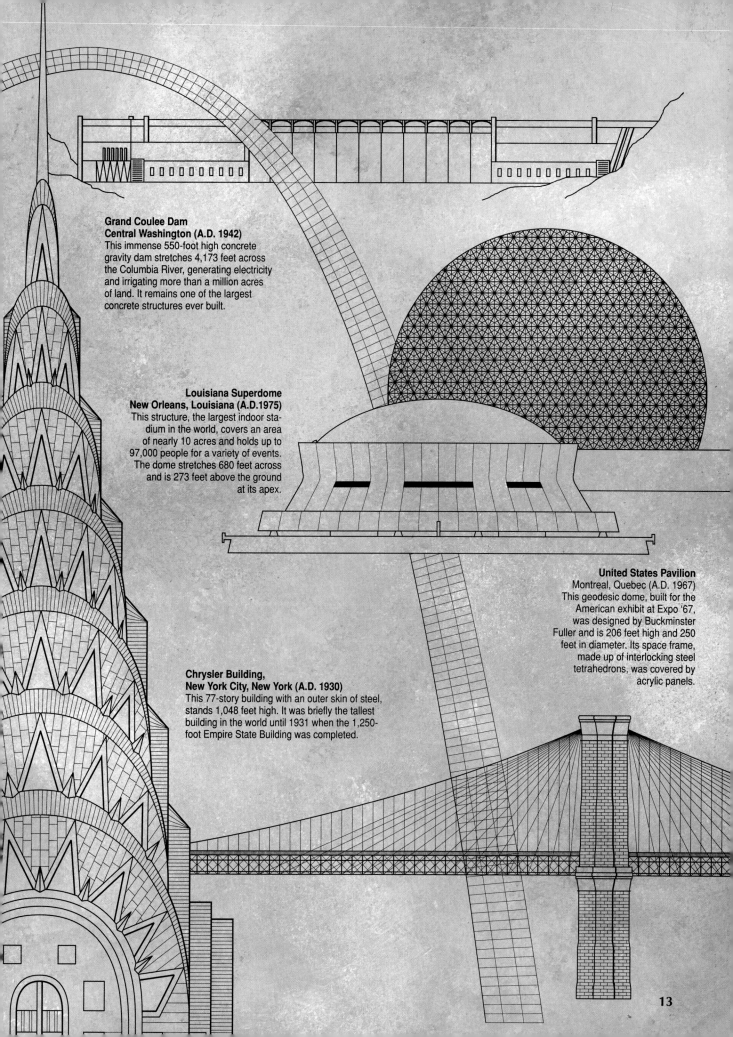

Grand Coulee Dam
Central Washington (A.D. 1942)
This immense 550-foot high concrete gravity dam stretches 4,173 feet across the Columbia River, generating electricity and irrigating more than a million acres of land. It remains one of the largest concrete structures ever built.

Louisiana Superdome
New Orleans, Louisiana (A.D.1975)
This structure, the largest indoor stadium in the world, covers an area of nearly 10 acres and holds up to 97,000 people for a variety of events. The dome stretches 680 feet across and is 273 feet above the ground at its apex.

United States Pavilion
Montreal, Quebec (A.D. 1967)
This geodesic dome, built for the American exhibit at Expo '67, was designed by Buckminster Fuller and is 206 feet high and 250 feet in diameter. Its space frame, made up of interlocking steel tetrahedrons, was covered by acrylic panels.

Chrysler Building,
New York City, New York (A.D. 1930)
This 77-story building with an outer skin of steel, stands 1,048 feet high. It was briefly the tallest building in the world until 1931 when the 1,250-foot Empire State Building was completed.

THE STRUCTURAL TUG OF WAR

t was the thick of rush hour in San Francisco, and Friday, October 17, 1989, seemed much like other days before it: The famed cable cars rumbled along on their usual clockwork crawl up and down the city's steep hills, while the steady flow of automobiles inched along the gleaming ribbon of the Golden Gate Bridge and among the concrete and steel towers that make up the city's skyline. Then, at 5:04 p.m., the earth shook. This was no mere tremor, not one of the gentle rumblings Californians have come to expect over the years. It was an earthquake of nearly the same intensity as the one that had ravaged the city eight decades earlier. Candlestick Park swayed. Skyscrapers squirmed on their foundations. Bridges wriggled in rhythm with the tremors. And millions of residents hung on as thousands of smaller structures shook. Finally, 15 frightening seconds after it began, it was over.

Although measuring 7.1 on the Richter Scale, the Loma Prieta earthquake was considered only a rehearsal for California's long-awaited "Big One." Nonetheless, its damage was incredible—buckled highways, pancaked buildings and collapsed bridges were scattered across the city and beyond. More incredible was the fact that most of the area's structures held their ground during the quake and escaped intact. Those that stood undamaged did so because the architects and engineers who designed and built them included enough safeguards to absorb the shock of a moderate earthquake.

Of course, it does not take a major earthquake to do serious damage to a work of civil engineering. Even the everyday forces such as the pull of gravity, the buffeting of wind and the wearing effects of water can combine to topple a poorly engineered building or collapse a badly constructed bridge.

Indeed, whatever their shape or function, all structures are engaged in an ongoing tug of war, with the external forces of nature on one side and the internal strength of the structure's design and materials on the other. Ideally, that strength will be sufficient to achieve a standoff with the many external forces that are allied

---

The columns of the ancient Temple of Aphaia, at Aegina in Greece, have endured all the forces that nature has thrown at them since the 5th Century B.C. Slightly the worse for wear, these powerful, well-engineered pillars continue to perform their structural duty.

against the structure. But any structure also needs enough additional strength in reserve—structural engineers call this a factor of safety—to accommodate any sudden, unexpected surges in nature's forces: an earthquake; a powerful gust of wind; or even an extreme increase in temperature. Conversely, any major flaw in a structure's design, the use of inferior building materials, or shoddy construction techniques can eventually lead to structural failure, even without the extreme forces of nature applying heavy loads.

In effect, all structures, from squat, unassuming warehouses and soaring bridges to tunnels far beneath the Earth's surface, must have the inherent ability to counteract the internal or external forces imposed upon them. These forces, commonly known as loads, can take many forms, but fall into two broad categories: static and dynamic loads. Static loads are further subdivided into dead and live loads. Dead loads include the entire weight of the structure itself, along with the weight of any permanently attached components, such as the walls, floors and ceilings of a building or the deck of a bridge. Live loads are the forces a structure bears through its use and under normal weather conditions. They change regularly and sometimes relatively quickly; people, furniture and stored materials are all considered live loads. So, too, are the cars and trucks that pass over a bridge or the rapid-transit trains that run through tunnels. Even the weight of rain, snow and ice on a structure or the force of flowing water against the piers of a bridge in a river or the ocean are live loads.

COMPRESSION: THE CRUSHING FORCE

Compression, sometimes called the crushing force, is the stress that pushes together and shortens material. It is present in virtually all structures, from ancient pyramids to modern skyscrapers. Compressive stress results first from the load of a structure's own weight, and then from the pressure of added loads. A standing column is always in compression. A beam placed atop it pushes down with even more force. If the column is made of weak material or the load is placed atop it too far to one side or the other, it will collapse. Concrete and rock, such as marble and granite, have long been the best materials to withstand compression. Today, both are surpassed by steel.

The micro-view
Compression is clearly visible at the microscopic level. The model above left shows the atoms of a hypothetical material in an unstressed state. The model above right shows how the atoms move closer together under compression, shortening and thickening the material. This reflects how all structures and materials compress.

Above-ground structures, particularly high, vulnerable ones, must deal with normal winds, one of the most important live loads. The pressure of the wind increases with the height of a structure since wind speed increases with altitude, and since it is striking a larger surface area. Wind, therefore, becomes a crucial factor in the design of skyscrapers and bridges. Indeed, the effect of wind on a tall skyscraper will be much greater than wind of the same velocity on a much shorter building. The influence of wind is further compounded by the fact that, depending on the shape of a building, wind can exert not one, but three kinds of force: pressure on the windward side of the structure, suction on the opposite side and turbulence on any side parallel to wind direction.

Structures that rely on many joints and connections also are subject to the force of thermal pressures brought on by shifting temperatures. Bridges with more than one section, for example, can expand on warm days and contract on cool days, straining the joints and creating weak links for the entire structure. Likewise, the concrete or steel comprising the dome of a covered stadium can expand or contract as temperatures fluctuate; if the temperature extremes are too great and the thousands of joints unable to cope, the dome could become unstable.

Dynamic loads, the other major category of forces, are those that have a sudden impact on a structure. If the ongoing structural tug of war is largely an even contest, dynamic loads are the sudden jerks one side might use against the other. The most dramatic are earthquakes and sudden gusts of powerful winds that sometimes strike

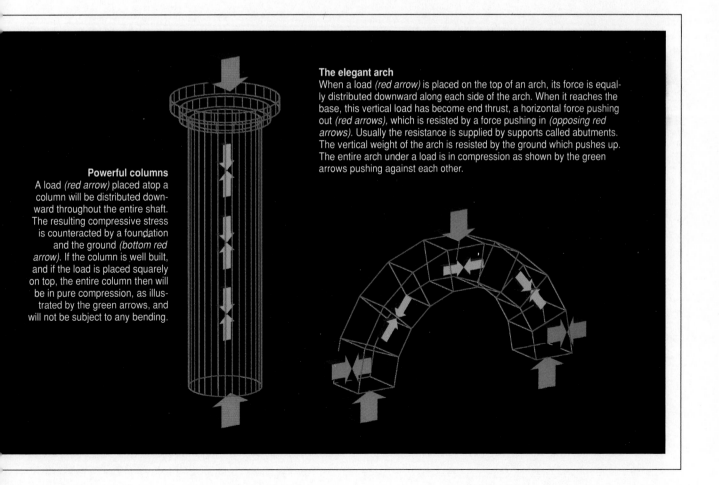

Powerful columns
A load *(red arrow)* placed atop a column will be distributed downward throughout the entire shaft. The resulting compressive stress is counteracted by a foundation and the ground *(bottom red arrow)*. If the column is well built, and if the load is placed squarely on top, the entire column then will be in pure compression, as illustrated by the green arrows, and will not be subject to any bending.

The elegant arch
When a load *(red arrow)* is placed on the top of an arch, its force is equally distributed downward along each side of the arch. When it reaches the base, this vertical load has become end thrust, a horizontal force pushing out *(red arrows),* which is resisted by a force pushing in *(opposing red arrows).* Usually the resistance is supplied by supports called abutments. The vertical weight of the arch is resisted by the ground which pushes up. The entire arch under a load is in compression as shown by the green arrows pushing against each other.

with disastrous effects. But dynamic loading also results from large ocean waves, sonic booms, the vibration of traffic or heavy machinery and even the bouncing effect of people walking along a floor. Often dynamic loads are unexpected and severe; occasionally they are bizarre. In 1945, a disoriented U.S. Air Force bomber hit the Empire State Building at a speed in excess of 250 miles per hour. There was extensive damage, but the solidly constructed riveted steel skeleton of the 365,000-ton building withstood the impact.

THE FIRST AND THIRD LAWS OF MOTION

Given what they are up against, it seems amazing that structures stand up at all. That they do is a reflection of Sir Isaac Newton's First and Third Laws of Motion, despite the obvious lack of motion exhibited by most large structures. These laws explain not only why objects move, but also why they do not move, thus providing the bedrock upon which structural engineering is based. They also explain why skyscrapers remain upright, how dams can hold back millions of tons of water, why bridges seem to hang suspended in midair, why tunnels can withstand tons of dead weight from above, and how domes cap vast inner spaces without crumpling. Newton's two laws of motion are so closely related as to be almost inseparable. And very often they are confused.

Newton's Third Law, known commonly as the law of action and reaction, describes how any two objects affect each other. It states that whenever one object

TENSION: THE PULLING FORCE

Known as the pulling force, tension is the stress that stretches materials. A rope, for instance, when it is not being pulled, is loose and flexible. A rope pulled taut during a tug of war, however, is in tension and becomes hard and firm. The rope will fail and pull completely apart if its material is weak or the combatants so strong that they pull its fibers apart. While tension has been exploited by builders for millennia—rope bridges and tents are both prehistoric tension structures—it was rarely used on a large scale because available materials lacked strength. Today, tension is the primary force at work in the strong steel cables of a suspension bridge, the inflated domes that cover some modern stadiums and the huge cable and acrylic tent-like roofs that are used to cover everything from airport terminals to playing fields.

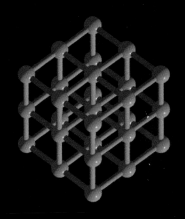

The micro-view
Above is a model of atoms in a hypothetical material under no stress. The model at right shows how atoms under tensile stress *(red arrows)* move further apart, lengthening and slightly narrowing. This process reflects the reaction of all materials in tension.

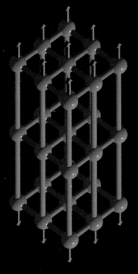

Steeling for stress
The steel cable shown at right is an example of a material placed under tension. When a pulling load *(red arrow)*, such as an elevator car, is placed on the end of the cable, the cable is pulled taut, giving it rigidity. The cable actually stretches under tension, as shown by the mauve arrows, but because it is made of woven steel fibers, it will withstand the stress.

exerts a force on a second, the second object must counteract with equal and opposite force. Forces are the result of the interaction of objects; every shove is met by a push. When a column pushes down on the ground beneath it, the ground pushes back with an equal and opposite force. What is true for the earth and a column applies to any two forces at work on a structure. The downward pressure of a concrete floor is resisted by the equivalent upward pressure of the horizontal steel girders that support it. The girders push down on columns below—which push back—and so it goes, right down to the ground.

The First Law, commonly called the law of inertia, describes the net effect of these many opposing forces at work against each other. It states that an object at rest, or moving at a constant speed in a straight line, will remain at rest, or continue to move at that same speed, unless it is influenced by other forces. A skyscraper, for example, is subject to a number of different forces—everything from the pull of gravity and the effects of unpredictable weather, to the way the building's own internal components act upon one another. However, if these forces acting on the skyscraper cancel each other out, the building remains anchored to the ground, standing in a state of equilibrium.

The difference between the two laws is simple, but subtle. Where the First Law determines how any structure will react under the influence of the many opposing forces acting on it, the Third Law dictates the rules under which very specific forces are transferred from one member to another.

STRESS AND STRAIN

Newton's First and Third laws underpin the calculations that engineers must make when they set out to determine how a structure and its myriad components will be affected by the many pressures, commonly called stresses, that will be brought to bear on it. Stress leads to strain, but the two are very different things. Stress is the amount of internal force exerted on an object; it describes, for example, how much pressure the concrete floor is putting on its supporting girder. Stress is always measured in terms of force and area—pounds per square inch (psi), for example. Strain, on the other hand, is the amount of deformation caused by the pressure; it might describe exactly how far a girder has bent or changed as a result of the stress from the floor. Strain is always expressed as a distance from the norm—the strain from pulling on a cable may be two inches. However, if and when stress leads to excessive strain, the crucial equilibrium is upset and then the structure might even topple or collapse.

Compression, tension and shear—pushing, pulling and sliding—are the most basic stresses. Under compression, an object is pushed together, thus shortening and thickening imperceptibly. Under the opposite stress, tension, an object is pulled apart, and becomes longer and narrower, or thinner. Shear occurs when an immovable object is pushed or pulled from opposite sides so that internal planes of the material slide across each other. In addition, structures often must endure variations and combinations of these three basic stresses: torsion, a twisting force that produces shear; bending, a combination of compression and tension in a single element; and buckling, the heavy compressive loading that causes spontaneous and extreme bending, and, inevitably, structural failure. Any or all of these stresses can be present in one structure at the same time. Engineers not only must anticipate

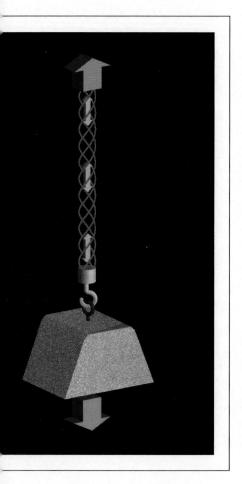

which forces will affect which elements of a structure—and when—but also must decide which materials and structural shapes will most successfully balance the ensuing stress and strain.

COMPRESSION

Certain shapes endure specific stresses better than others. These fortuitous relationships are reflected in the history of structures, and are as important for the people who construct office towers in modern Chicago as they were for temple builders of ancient Rome. In dealing with compression, for instance, the oldest and most effective structures are columns, which—along with beams—are the building blocks for everything from doghouses to skyscrapers.

With a beam or floor pushing down on it, a well-made column transmits the heavy load vertically through its entire volume to a foundation or another compressive structure below. Here, a principle called the middle third rule is important. The column must be engineered so that the center of the load is applied to the middle third of the top of the column. When the load is centered on this middle third, the entire column is in compression. If the load is applied too far to one side or the other—or if the load is simply too great for the column—the column will begin to bend. When an external force on a column becomes so great that it turns into a critical load—the point at which it has been strained too far—the column buckles. Unpredictable and sudden, buckling is a fast-moving deformation. The roof of the Hartford, Connecticut, Civic Center's hockey rink took just 10 seconds to collapse in 1978 after a few of the roof's compressive steel support rods buckled, causing other rods to follow suit in a domino-like progression. The middle third rule applies to any structures in compression, from the walls of buildings, to bridge piers and massive concrete dams.

The age-old problem with the column and beam combination is that many vertical members are needed to support large and heavy horizontal elements that will sag without closely spaced vertical supports. Though the column and beam system worked well for small, roofed structures, it proved inefficient for edifices, such as the Parthenon, atop the Acropolis in Athens, that had long unsupported beams and a massive roof. Such a structure needed many dozens of columns to support the enormous load from above. Fortunately, engineers have long had at their disposal a solution in a shape that is as graceful as the most graceful column, able to span openings as narrow as a passageway or as wide as a river and that is stronger than the strongest beam: the arch.

Builders have been obsessed with the arch for centuries. The graceful shape probably originated in ancient Egypt or Mesopotamia more than 5,000 years ago. Nevertheless, it took the ancient Greeks and, to a much greater extent, the Romans to realize the arch's potential. In their hands, arches became the centerpieces of triumphal entry gates and the heart of amphitheaters, bridges and aqueducts from Italy to England. Later generations of builders took new materials—successively concrete, iron and steel—and used the arch to span spaces that would have dumbfounded their predecessors.

An arch conducts the downward force of compression through its curving form toward the ground. Along the way, vertical compression becomes a horizontal force, known as end thrust, which must be borne at the feet of the arch by strong

SHEAR: THE SLIDING FORCE

When a material is pushed from opposite sides externally, it is forced in opposite directions internally. The resultant stress is called shear, the sliding force. The blades of scissors, as they pass by each other, cut through paper by forcing it in opposite directions. In structures such as buildings, this stress is commonly withstood. Columns push upward on a beam, but the beam sags in the middle. Shear stress forms at the border between the opposing upward and downward pushing forces. As long as the beam is made of sufficiently strong material, such as steel, it will withstand the shear stress.

supports. Firmly anchored in the ground, these supports absorb and redirect the force, allowing it to flow into the surrounding earth. In large arched structures, the supports required must be massive and are commonly known as abutments. They, too, absorb the horizontal force of end thrust and transmit it into the ground, hence many arched bridges and dams are built into hard, rocky canyon walls or mountainsides. Even a simple arched doorway will collapse if end thrust is not absorbed by supporting walls; in the case of a Gothic cathedral, the job is done by huge buttresses that safely channel the outward thrust of the arch.

Traditionally, stone was the best material to withstand the thousands of pounds per square inch that compression can bring to bear. For millennia, the compressive strength of marble, granite and limestone were unmatched by any other building material. Even wood, despite its long history as a building material, has only a fraction of the compressive strength of stone.

Man-made stone, too, has been dealing successfully with compression ever since ancient builders first stirred water into lime to form a primitive cement that bonded sand and stone into rock-hard concrete. In its present-day form, concrete can be expected to handle compression of 3,000 to 5,000 psi; specialized high-strength concrete can withstand anywhere from 6,000 to 20,000 psi. But even that pales in comparison to modern steels, which are manufactured to withstand compression and tension of up to 36,000 psi. Steel, however, is much heavier and more expensive to manufacture than concrete. Moreover, it is susceptible to

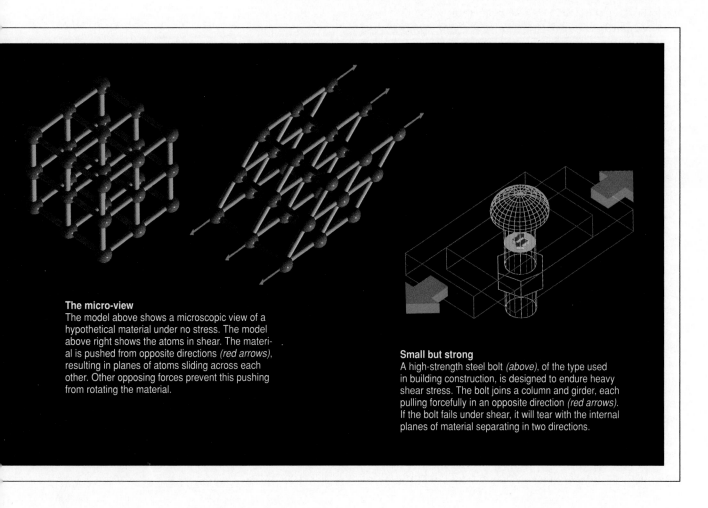

The micro-view
The model above shows a microscopic view of a hypothetical material under no stress. The model above right shows the atoms in shear. The material is pushed from opposite directions *(red arrows)*, resulting in planes of atoms sliding across each other. Other opposing forces prevent this pushing from rotating the material.

Small but strong
A high-strength steel bolt *(above)*, of the type used in building construction, is designed to endure heavy shear stress. The bolt joins a column and girder, each pulling forcefully in an opposite direction *(red arrows)*. If the bolt fails under shear, it will tear with the internal planes of material separating in two directions.

extremes in temperature, becoming more brittle when the thermometer plunges below -30°F, and losing much of its strength in fire of more than 1,200°F. In such extreme conditions, concrete, however, holds firm.

TENSION

The opposite stress to the pushing of compression is the pulling of tension, a force well illustrated by the stretching of ordinary rubber bands or the steel cables of an elevator pulled taut by the weight of its cab and passengers. Tension is the driving force behind a suspension bridge, at the heart of which hang the signature tensile cable spans that mirror a compressive arch bridge in form and function. Where the concrete or steel arch of a bridge channels compressive weight from above, an inverted arch—formed by cables strung between two large towers—absorbs tension caused by weight pulling from below. And where an arched bridge relies on abutments to withstand massive pushing, a suspension bridge relies on similar structures to counteract a massive pulling force. In this case, the cables are attached to huge concrete blocks, called anchorages, on either bank. Decks of suspension bridges are also held in place by vertical suspenders that hang in tension from the main cables above.

Originally, natural vines and woven ropes were the materials used for tensile structures such as bridges. Later, iron chains did the job. But now, bundles of steel wire, which make up the cables of a suspension bridge, are by far the strongest material in tension. Such is the strength of steel that thin wires made of newer kinds of high-tensile steel can tolerate stresses of 300,000 pounds per square inch, enough support to dangle the entire Leaning Tower of Pisa, like a gargantuan pendant, from a cable barely more than an inch in diameter.

Despite its great tensile strength, steel is not immune to deformation under tension; a heavy enough load could stress a steel cable to its yield point—a structural point of no return, beyond which steel loses its natural elasticity and undergoes irreversible deformation. Fortunately, this quality can be used to a structural advantage. Paradoxically, controlled stretching beyond a certain point during the manufacturing process leads to a stress-hardened material, called cold-deformed steel, that renders the material brittle but gives it greater strength. Such stress-hardened steel is used for the vertical suspenders that connect a suspension bridge's heavy deck to its huge suspended cables.

Affordable and efficient alternatives to steel cable seem to be frustratingly few. Wood, which is reasonably strong in compression, also endures tension well. But it is somewhat weaker when the stress is applied against the run of the grain, and even at its strongest is no match for the tensile strength of steel. A cable of aluminum alloy, on the other hand, has all the tensile strength of a steel cable and has only one-third of the weight. But aluminum alloy is prohibitively expensive.

By comparison, concrete, which is immensely strong under compression, is a weakling under tension. In fact, with a tensile strength just one-tenth of its compressive strength, concrete was for centuries of little use in structures subject to tension—and would be still but for the development in the late-1800s of a new building material that married the compressive strength of concrete to the tensile strength of steel. Reinforced concrete beams and columns are made by embedding steel bars, or re-bars, in the concrete before it hardens. Usually placed near the bottom of a beam or up the side of a column that would otherwise bow, the steel skeleton imbues the concrete with a much greater resistance to the tension that develops when bending takes place in long, unsupported beams. Mats of pre-welded wire mesh are embedded similarly to reinforce large concrete slabs for floors or to create eggshell-thin concrete membranes used to roof structures that, when finished, appear to have been sculpted rather than built.

Beams, like the columns that support them, are also put into compression under load. But because of the combined effect of the downward load and the upward push exerted by the columns on its ends, a beam also develops tension. This combination of forces causes the beam to bend, as its upper surface shortens under the stress of compression and its lower surface lengthens due to tensile stress. Bending is by no means a catastrophic situation; all beams will bend to a predetermined safe degree. The goal of builders is to use beams made of materials that can endure bending with a limited number of columns for support. Traditionally, wood has been one of the best materials for beams; stone and old concrete required regular and closely spaced supports. Even today, wood is still commonly used, but steel beams can endure considerably more bending stress than any other construction material.

SHEAR STRESS

Regardless of the material or the shape, all structures are susceptible to shear, the sliding stress. Beams are especially prone. The resulting stress causes one part of a structural element to slide past an adjacent part in much the same way as scissor blades cut paper when they slide by each other. A beam supported on its ends by two columns can develop shear at the points where the horizontal member meets the vertical ones. Since the ends of the beam, and the middle of it, are simultaneously forced in opposite directions, particles within the beam are also pushed from opposite directions past one another. If the stress is too great, the beam can be ripped or sheared apart, usually near the support. The most impressive example of resistance to shear is seen in the high-strength bolts used to join the steel frames of skyscrapers. The normal opposing forces brought to bear on bolts by the steel columns and girders they join are immense; when the building begins to move in the wind, they must endure many times more stress.

There is no best shape for resisting shear; withstanding the stress successfully is entirely dependent on materials. Wood (when the pressure is not parallel to the grain), steel and reinforced concrete are, to varying degrees, strong under the pressure of shear, while ordinary concrete is particularly weak.

Torsion is a type of shear stress that occurs when a structure is subjected to twisting force. A skyscraper, for example, is actually a giant vertical column projecting from its immovable foundation, but unattached at the top. The sideswipe of a strong wind, or a quake's shaking, can cause the entire skyscraper to twist in torsion. Huge towers are designed with devices to combat these stresses—from exterior steel cross-bracing that stiffens the building, to huge computer-controlled blocks on the roof that move to counteract the forces. When designed to handle such stress, a building should give, but return to its original position.

Some structural shapes are more effective in torsion than others. A hollow tube, whether rectangular, circular or square, is more efficient at resisting torsion than a solid bar or beam of the same weight. The more tubes the better. This principle can be applied to skyscrapers that may be subject to earthquakes or high winds. Viewed from the outside, for example, Chicago's 1,450-foot-tall Sears Tower looks like an ordinary rectangular skyscraper. Yet a cross-sectional look at the tower reveals that it is, in fact, nine square tubes whose interior walls are braced against one another. Known as bundled tubes, these provide more resistance to torsion than a single large tube would. At the same time, the shared interior walls add stiffness to the overall structure and help keep the high exterior walls from shearing or buckling.

THE POWER OF THE TRIANGLE

If the arch is an enduring shape from the architectural past, then the triangle may be the form of the present and future since it provides the basis for the ubiquitous truss. Tension and compression can be used to a builder's advantage if various structural members are combined to form a number of triangles arranged in a framework called a truss. The strength of a truss—and its simple beauty—is based on a simple geometric fact: A triangle is the only shape that cannot be deformed without changing the length of one of its sides. If, for example, a hinged square frame is pushed, its angles and shape will change, but it will remain intact. On the other hand, if a hinged triangular frame is pushed, it does not move, because its angles and shape cannot change. The same principle applies for a series of con-

FRUITFUL COMBINATIONS

When two or more of the basic stresses are present in a structure, they often combine to form new loads. An off-center vertical load will create a bending stress in a column—one side of the column is in tension, the other in compression. Although this is not desirable,

if the load is not too extreme, the column will withstand bending. Beams, on the other hand, are expected to bend slightly under loads and are intentionally built of materials strong in both tension and compression. A combination of forces is also at work in trusses,

the simple, strong frames made up of triangles. However, each separate segment of a truss is in either pure tension or pure compression; none of the segments bend on their own. Despite this, like the beams they often replace, trusses do bend as a whole.

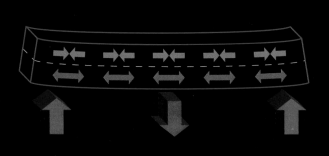

Bending beam
Beams, such as the one at left, are intended to endure bending. A beam supported by two columns (end arrows) sags slightly in the middle when enduring a dead load (middle arrow). The side of the beam above the central line or neutral axis shortens in compression. The side below the neutral axis lengthens in tension. The axis itself, in neither tension nor compression, stays the same length. Steel, aluminum, reinforced concrete and wood are the best materials to endure bending.

nected triangles that constitute a truss; it also cannot be deformed by its own dead load or by a live load. Instead, each of its members is designed to push or pull against another. An ideal truss owes its remarkable strength-to-weight ratio to the fact that all its members are in either pure tension or pure compression. Ideally, none undergo bending. Trusses use less material, are lighter and cheaper than solid structures of the same size—and they carry the same loads. Indeed the truss is a structural testament to the old adage that less is sometimes more.

Traditionally, trusses have been made from wood to take advantage of the material's natural tendency to resist both tension and compression. Wood still finds its way into many modern-day trusses, especially in residential construction, where prefabricated trusses are used to support roofs. In larger construction projects, steel trusses, with greater strength and rigidity, and reinforced concrete trusses have largely supplanted their wooden counterparts. These powerful sets of triangles have long been used to support the decks of bridges and wide roofs. Today's steel-truss bridges are descendants of the timber-truss bridges, including that icon of colonial America, the covered bridge, whose barnboard siding masks a backbone of truss-work. With the coming of the railroad in the 1800s, the wooden truss also stiffened the spines of hundreds of seemingly rickety, but deceptively strong, railway trestles that allowed the transcontinental trains to cross some of the world's most rugged terrain. In this century the truss has undergone a renaissance and is now used in everything from fabric tents and geodesic domes to revolutionary skyscrapers.

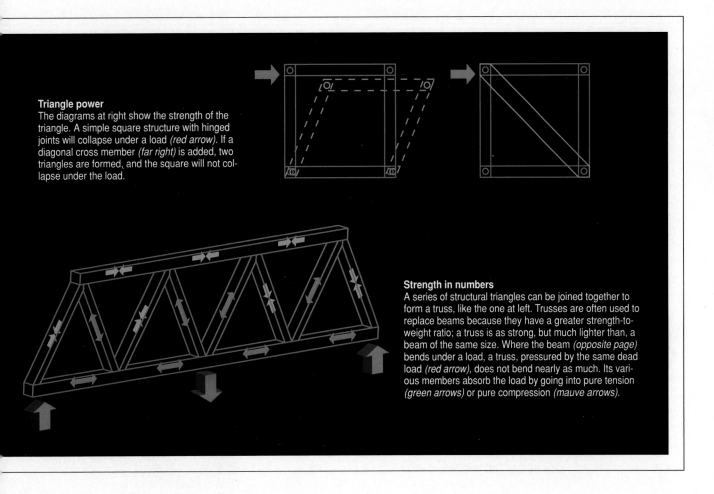

Triangle power
The diagrams at right show the strength of the triangle. A simple square structure with hinged joints will collapse under a load *(red arrow)*. If a diagonal cross member *(far right)* is added, two triangles are formed, and the square will not collapse under the load.

Strength in numbers
A series of structural triangles can be joined together to form a truss, like the one at left. Trusses are often used to replace beams because they have a greater strength-to-weight ratio; a truss is as strong, but much lighter than, a beam of the same size. Where the beam *(opposite page)* bends under a load, a truss, pressured by the same dead load *(red arrow)*, does not bend nearly as much. Its various members absorb the load by going into pure tension *(green arrows)* or pure compression *(mauve arrows)*.

BRIDGES

Presiding over the entrance to New York Harbor, the Verrazano-Narrows Bridge is a perfect marriage of strength and beauty, fulfilling the goal of its designer, Othmar Ammann, to build an "enormous object, drawn as faintly as possible." Even drawn faintly, the mighty suspension bridge looms large. For many years after its completion in 1964, it reigned as the world's longest span, its 70-story-high twin towers separated by almost a mile, a span so great that the curvature of the Earth was factored into the bridge's design. Equally impressive are the 7,200-foot-long steel cables. It took six months just to "spin" the four massive main cables; workers unspooled tens of thousands of individual, pencil-thin wires and then compressed them into three-foot-thick bundles. Laid end to end, these separate threads of wire would stretch 143,000 miles—halfway to the Moon. The pull of the bridge's main cables exerts a quarter of a billion pounds of force on the anchorage blocks to which they are attached. The blocks themselves stand 10 stories tall and weigh almost 400,000 tons each. Even the bridge's paint job offers awesome numbers: The structure requires 36,725 gallons for each repainting.

For all its immensity, the Verrazano-Narrows Bridge works on the same structural principles that have governed simple rope bridges since human beings decided that it was necessary to get to the other side of rivers and gorges. Indeed, all the grand—and not so grand—bridges of today, whatever their design, have their antecedents in ancient spans. A concrete overpass spanning a modern superhighway works essentially the same way as a felled log bridges a stream. And a Roman arch builder would understand instantly how a 1,000-foot steel arch bridge soars, seemingly unsupported, across a harbor. What is new are the improved understanding of the physical principles of bridges, and the building materials and construction technologies in use. The result is that modern bridges carry heavier loads, span ever-longer distances and use less material to do the job.

The Maurice J. Tobin Memorial Bridge, a cantilever span, soars across the Mystic River as it flows into the Boston Harbor. The bridge finds support for its massive weight in powerful foundations beneath the water surface, and develops rigidity to combat the stresses from use and weather through a system of diagonal steel trusses.

BRIDGE TYPES

At right are three of the most common bridge types—cantilever *(top)*, arch *(middle)* and suspension *(bottom)*. The black arrows represent the load of the bridge; the red arrows show how the load is counteracted. A cantilever bridge, a variation of a beam bridge, normally is composed of three sections: two arms that extend from opposite shores and a connecting span in the middle. The weight of the structure is borne by the piers and their foundations; abutments at each end anchor the bridge to the shores. The arch bridge channels its weight down along the sides of the arch where it is counteracted by onshore abutments buried in the ground. The tendency of the abutments to move outward is counteracted by the solid ground surrounding them. In suspension bridges, the deck is suspended from tensile steel cables that are draped between high towers and secured to onshore anchorages that counteract the pull of the cables.

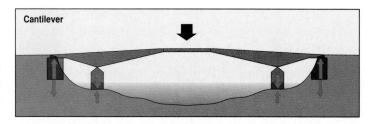

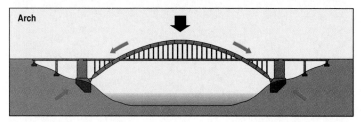

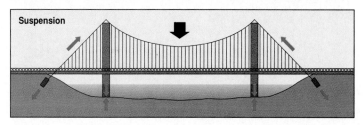

A TRIUMVIRATE OF SPANS

Despite the diversity of spans throughout the world, bridges are divided into three basic types—beam (which includes cantilever), arch and suspension—each defined by the way it bears the loads imposed by its weight, its use and the forces of nature. Though many factors, including economics, existing technologies and even aesthetics, contribute to the decision of exactly which type of structure to build, the most important consideration is always related to the site of the bridge—the distance to be spanned, the type of terrain on which the structure will stand and the day-to-day traffic to be carried by the bridge.

The simplest and oldest of all the bridge types is a horizontal beam supported either by firm ground or by columns, known as piers, on both banks of a river or gap. Piers, which must stand on a solid foundation, bear the brunt of the dead load of a bridge. The deck, in essence a beam, also called a girder, bends between the piers—but only slightly. The size of the beam is selected so that sagging is barely noticeable. In a more complicated variation of the simple beam span, the cantilevered bridge, the deck is supported in a different way. Where the deck in a simple beam bridge is held up at each end, in the cantilever, beams are anchored at each bank, then supported by piers partway across the river or gap. When one end of a beam is firmly attached to a solid anchorage on the shore, the other end does not totter under a load; moreover the free end can extend a considerable distance unsupported, much like a diving board. A cantilever bridge comprises two such arms, one extending from each shore; the arms are joined together by another smaller beam.

Arch bridges, the second main bridge type, obtain their strength and spanning ability from the compressive arch itself, which can extend long distances without the need for supporting piers. The decks of such bridges may rest atop the arch, partway down the arch, or at the bottom of the structure. Whatever the position of the deck, the weight of the bridge is channeled down the arch into massive support abutments at its lower ends.

Suspension bridges are the longest spans in the world because they take advantage of both the strength and relative lightness of steel cable. In a suspension bridge, the deck is not supported by a compressive arch, but by an enormous set of steel-wire cables strung over two high towers. The deck hangs from this graceful twin-peaked span by vertical steel cords known as suspenders; the resulting load puts the powerful overhead cables in tension. Heavy, onshore anchorages withstand the resulting pull on the main cables, while the towers work in compression to support the weight of the entire bridge.

THE FORCES AT WORK ON A BRIDGE

All bridges, regardless of their type, must endure a variety of natural and man-made forces that exert dead, live and dynamic loads. The dead load is the weight of the bridge; live loads include the movement of traffic across the deck, and normal winds and weather, including snow and rain; dynamic loads are gusts of wind or earthquakes that oscillate the bridge. The longer a bridge is, the greater its dead and live loads will be. Most bridges are founded on bedrock—the solid rock that underlies soil and broken rock soil—and necessarily so, for large bridges weigh hundreds of thousands of tons.

The effects of weather are also critical. A bridge is usually an exposed structure, built high and with little shelter afforded by the surrounding terrain; even moderate winds can sway and oscillate both flexible structures such as suspension spans and rigid structures such as cantilever bridges. Daily fluctuations in temperature can also produce live loads, since many building materials expand when heated

This photo shows an on-screen view of a railway bridge superstructure being stress tested by a specialized computer program. Engineers use such programs to bypass laborious model building and testing. At this point, the bridge is only under the dead load of its own weight. The color of each segment corresponds to the relative level of stress that the part of the bridge endures. The on-screen working tools at the right include a color key at the bottom for identifying stress levels: from blue, which indicates no stress, to red, which signals the highest amount of stress.

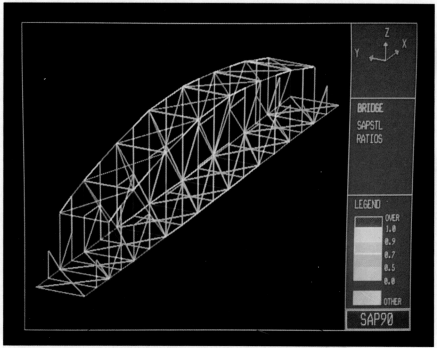

and contract when cooled. This vulnerability to temperature proved to be both a bane and the salvation of the Forth Railway Bridge, the mile-and-a-half-long cantilever spanning Scotland's Firth of Forth. During its final construction phase in 1889, icy winds cooled the steel to such an extent that the final joining span would not quite meet the section stretching out from the pier. The ingenious chief engineer placed naphtha-soaked wood shavings for 65 feet along the adjoining sections and set the shavings ablaze. As he had anticipated, the heat gradually expanded the steel enough for the two sections to be bolted together. In northern climes, bridges often must endure a spin-off effect of weather: Road deicing salt is used to melt snow, decreasing the load, but the salt is extremely corrosive and can gnaw through parts made of steel, weakening them.

The dynamic load of gusting winds can be an engineer's nightmare and has been responsible for a number of notorious disasters. Among these was the 1854 collapse of a suspension bridge in Wheeling, West Virginia, which, in the words of one eyewitness, went down in high winds with "an appalling crash and roar." Almost a century later, in 1940, the four-month-old Tacoma Narrows Bridge, known to motorists as "Galloping Gertie" because of its extreme and dangerous flexibility, suffered a similar fate. The design of the bridge's deck was also flawed, its nonaerodynamic shape causing it to undulate in moderate winds. On the morning of November 7th, though the wind was only 42 miles per hour, the undulations became violent oscillations; large sections of the bridge tore off and fell into the narrows below. Fortunately, there was sufficient time to clear all traffic from the deck and no lives were lost.

Earthquakes, with their potentially lethal dynamic loads, are the enemies of all bridges, though some designs withstand the tremors better than others. Suspension bridges, with their inherent flexibility, tend to endure seismic shocks better than all other bridge types. For example, in the 1989 San Francisco earthquake, the Golden Gate Bridge, a suspension span, came through virtually unscathed. The bridge oscillated for about a minute—four times longer than the quake itself—but gradually came to rest, undamaged. On the other hand, the Oakland Bay Bridge, a complex of cantilever and suspension spans, suffered severe damage at some deck joints, though it did not collapse. While the flexible suspension spans remained unharmed, a portion of the simple beam approach to one of the cantilever spans did collapse. One older elevated highway in the area also failed and plunged to the ground.

Still another kind of natural phenomenon, known as scour, must be endured by bridges with piers that are founded in water. Scour occurs when flowing water erodes riverbeds and banks; this removal of sediments undercuts piers. Worse, it remains an invisible threat unless it is detected by inspection. Fortunately, engineers have many tools and processes by which to ensure that a bridge will withstand these loads, stresses and problems.

WIND TUNNELS AND COMPUTERS

Like all structural engineers, bridge specialists use standardized charts and tables to calculate dead loads, weight-bearing maximums for specific construction materials, live loads and dynamic wind stresses on certain shapes and materials. Then they calculate in an all-important safety factor. Often, engineers have learned from

Suspender
Hanging vertically from a main cable, these steel ropes support the deck.

Deck
Hangs from suspenders and is secured to the towers; on other bridge types, the deck is supported by piers.

Saddle
Saddle-shaped and of cast steel, a saddle atop each leg of a tower cradles a main cable.

ANATOMY OF A BRIDGE

No two bridges are identical; each one is engineered to provide the most dead, live and dynamic load-bearing strength possible under a specific set of conditions. A suspension bridge, such as the one below, is optimized to span long distances, but nevertheless has basic features in common with all bridges. Its foundations bear the weight of the bridge and must be strong in compression; the towers channel the bridge's weight to the foundations. The deck, along which traffic travels, must be sufficiently rigid to avoid dangerous oscillations. Finally, the bridge must be made of durable materials—normally steel and concrete—that can withstand the heavy loads of natural and man-made forces.

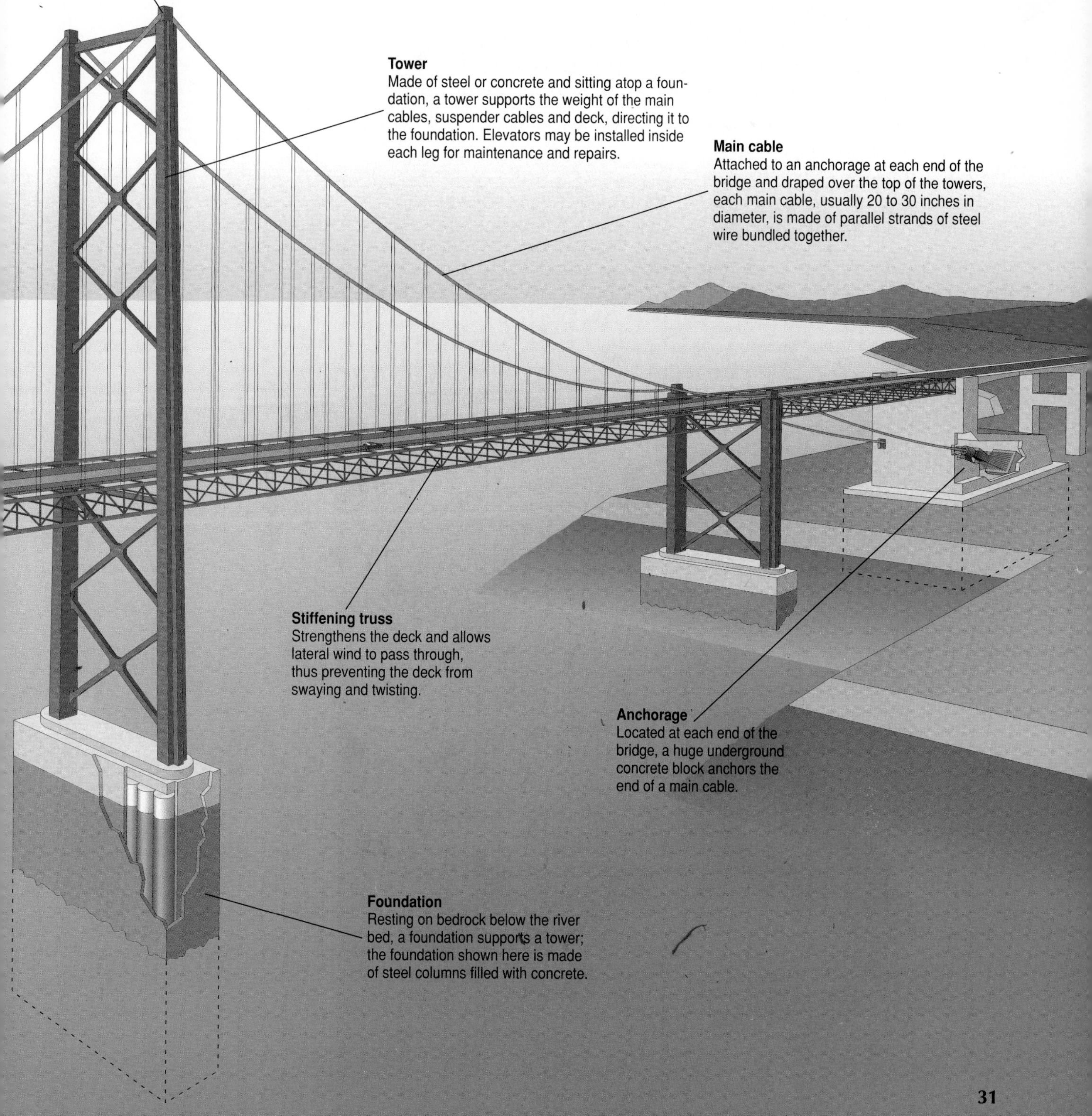

Tower
Made of steel or concrete and sitting atop a foundation, a tower supports the weight of the main cables, suspender cables and deck, directing it to the foundation. Elevators may be installed inside each leg for maintenance and repairs.

Main cable
Attached to an anchorage at each end of the bridge and draped over the top of the towers, each main cable, usually 20 to 30 inches in diameter, is made of parallel strands of steel wire bundled together.

Stiffening truss
Strengthens the deck and allows lateral wind to pass through, thus preventing the deck from swaying and twisting.

Anchorage
Located at each end of the bridge, a huge underground concrete block anchors the end of a main cable.

Foundation
Resting on bedrock below the river bed, a foundation supports a tower; the foundation shown here is made of steel columns filled with concrete.

disaster. After the collapse in high winds of the Tay Bridge across the Scottish Firth of Tay took 75 train passengers to their deaths in 1879, an inquiry revealed that its designer and chief engineer, Sir Thomas Bouch, vastly underestimated wind as a significant load. Not surprisingly, Bouch was removed from his next project—the Forth Bridge. And after the writhing death of the Tacoma Narrows Bridge, engineers began to design decks stiffened by open trussing, which also would allow winds to pass through instead of presenting a solid profile in the face of the aerodynamic forces. The disaster pinpointed the need for aerodynamic research on bridges and resulted in a better understanding of the behavior of flexible structures exposed to wind.

No bridge design today leaves the engineering office without first undergoing wind-tunnel testing and computerized stress analysis. A scale model of all or part of a bridge is built and installed in the test chamber of a wind tunnel—a device that generates a controlled flow of air in an enclosed environment. The results

Suspensions of Silk

Silk is ejected from the spigot, or silk gland, at the end of the spider's abdomen. Four to six spinnerets, seen here surrounding the spigot, distribute the material as it is ejected and weave the filaments together to give it composite strength.

In one of nature's most impressive paradoxes, the gossamer stuff of silken spider webs is one of the world's strongest materials. The only man-made material that comes close to matching the strength of a spider's silk is steel and the strongest silk has a tensile strength five times more powerful than its equivalent weight in steel. Perhaps less surprising, then, is that when a spider spins a web, it is actually building a suspension bridge. Like its man-made counterpart, the main "cable" is strung between anchored supports. But where a stiffened deck hangs from the cables of a steel suspension bridge, a spider's web hangs from the overhead silken thread. Spiders produce their silk in special glands, different glands producing different types of silk. The supporting silk that forms the outer web, for example, is tough and relatively nonelastic, while the silk of the inner web is stickier—to hold the spider's catch—and more elastic—to cope with the force of the wind.

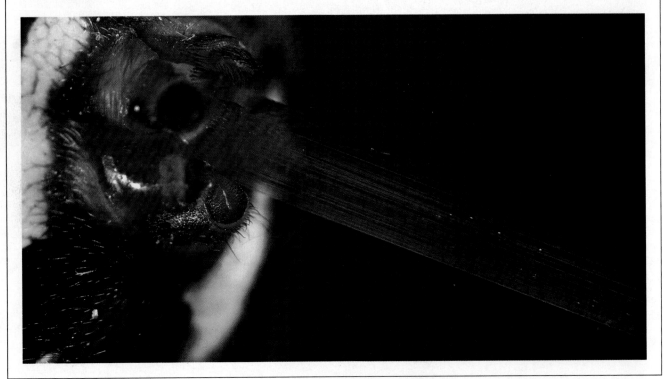

of such tests provide an accurate indication of a bridge's performance under various wind loads, and allow engineers to fine-tune their designs for the most aerodynamically efficient use of materials. This often can lead to substantial financial savings. Wind-tunnel tests on models played an important role in the construction of the Annacis Island Bridge in Vancouver, Canada. The tests showed that aerodynamic stability for the bridge would be possible with 20 per cent less structural steel than called for in the original plans. Moreover, preliminary tests convinced the designers to move girders to more efficient structural positions and to arrange the bridge's sidewalks differently than planned. Wind-tunnel testing also is used to assess bridges that were built before such testing came into use. Models of already-existing structures can be put through tests and if results warrant, structural modifications can be carried out.

Computer-aided design, or CAD, allows for a similar streamlining of the design and construction of bridges, and bypasses traditional draftsmen. With the appropriate software, a computer can digest any combination of specifications and generate alternative designs in a range of structural materials, complete with sets of cost estimates. In addition, computers allow engineers to isolate any single part of a design and calculate the stresses and strains to which that component might be subjected under various loads. And a CAD program provides use of a data base to update and adjust automatically all the elements of a bridge's plans and specifications. This saves enormous amounts of time. Thirty years ago, the engineers for the Verrazano-Narrows Bridge generated tens of thousands of technical drawings; today, these could all be computer-generated, stored on disk and each quickly brought to the screen at will for modification. Just as important is the sheer processing speed that modern computers offer, reducing to a fraction the amount of time that formerly was spent on the detailed and laborious calculations that had to be performed long before construction could begin.

NEW MATERIALS

Today's long-spanning bridges would never have been possible without the continuing developments in steel-making technology that began during the middle of the last century. The most common construction metals had been wrought iron, which was supple, but soft, and cast iron which was hard, but brittle. Steel was stronger than either type of metal and retained the best characteristics of both, making it strong in both tension and compression.

Artisans had made steel in limited batches for centuries, of course, but it was expensive and the process was part alchemy and part metallurgy. Rapid change in bridge design and construction came with the mass production of high-grade, low-cost structural steel during the middle of the 19th Century. In practical application, this meant that long and relatively slim girders of steel could now be manufactured to withstand enormous stresses. Moreover, steel's multipurpose characteristics meant that it could also be used for the construction of larger trusses, and steel columns and towers. Even so, it was some time before the new structural steel won the full trust of engineers. In England, in fact, the use of steel for bridge construction was prohibited until 1877. But in America, many unbelievers were firmly convinced when James Buchanan Eads completed a 1,500-foot-long bridge of three connected steel arches over the Mississippi River at St. Louis, Missouri, in 1874. The three arches, each more than 500 feet in length, constituted the first large steel crossing in the United States. Improvements in the stress-bearing qualities and mass production techniques of steel continued unabated during the 20th Century with research focusing on producing an even stronger steel and one that was more resistant to corrosion.

The other main material of bridge construction—concrete—has, on the other hand, long been important, but until recently was used almost exclusively as the stuff of bridge foundations. During the 19th and early 20th Centuries, two important developments occurred. The first was a better understanding of the chemistry of concrete and of the best mixing ratios for aggregates and cements. This meant not only that concretes could withstand more compressive stress and bear the dead load of enormous bridges, but also that engineers could more accurately predict what that bearing strength would be. With its great and predictable strength under compression, concrete was ideal not only for foundations and piers, but also for arch bridges. Furthermore, it was much cheaper to manufacture than steel. But concrete is weak in tension and was largely useless for girders and decks—which must endure bending—until a second development occurred. A whole new area of use unfolded at the end of the 19th Century when engineers learned how to combine steel and concrete to produce reinforced and prestressed concrete. When steel rods are embedded in concrete, forming reinforced concrete, the material gains tensile strength. When the steel rods are stretched, or prestressed, and attached to the ends of the concrete, the concrete becomes compressed before a load is applied, and is known as prestressed concrete. This material is not only strong in tension, but it absorbs loads with less bending than simple reinforced concrete. Reinforced and prestressed concretes are true composite materials because the steel and concrete respond as a single unit.

STURDY FOUNDATIONS

These new understandings of materials led to the construction of increasingly long-spanning bridges, playing important roles in every aspect of these structures. The stability of a bridge's superstructure—the deck—depends to a large extent on a sound substructure below: the abutments for an arch, the anchorages for a suspension bridge and any piers for these spans. The foundations, piers and abutments bear the immense load of the superstructure, which includes the deck, stiffening trusses, and overhead cables and towers. This weight can reach astonishing figures: The deck, cables and towers of the Verrazano-Narrows suspension bridge, for example, weigh about 187,000 tons and the pressure exerted by Australia's Sydney Harbour arch bridge on its abutments is about 88,000 tons.

One of two basic types of foundations is used for a bridge pier: massive or pile. The massive, or block, foundation is a concrete structure that channels loads directly into the surrounding ground. Ideally, it will be founded on a solid, load-bearing geological strata—preferably bedrock—and its sole duty is to be strong in compression. This base is essentially a huge concrete block or pillar that sits atop the bedrock and extends to the surface of the water. However, solid rock often may be located some distance below the ocean floor or riverbed, as is the case for the Verrazano-Narrows Bridge, with its massive reinforced concrete monoliths supporting two towers. Each monolith is 229 feet long and 129 feet wide. The footing

The Forth Railway Bridge, which crosses the Firth of Forth in Scotland, is a classic example of a steel cantilever bridge. Completed in 1890, the Forth Bridge comprises two connected spans, carefully balanced on three massive concrete piers, and stiffened by an elaborate system of trusses and steel stays. The bridge carries two railway tracks 150 feet above the firth and is still in active use today. At the time of its completion, it was the largest steel structure in the world, and contained the longest unsupported span—more than 1,700 feet. It is still considered one of the great bridge-engineering achievements.

on the Brooklyn end penetrates 170 feet—about 17 stories—beneath the water until it reaches solid ground; the Staten Island-end footing goes down only 105 feet—a mere 10 stories—below the water.

Where the bedrock is buried too deeply to be reached by digging or where the surrounding soil is unstable, the bridge piers can rest on a pile foundation, a cluster of specially treated wood, steel or concrete piles that have been driven deep into the ground. The Pontchartrain Bridge in Louisiana sits on foundations made of prestressed concrete tubes driven into the lake bed. Each pier of the Narrows Bridge in Perth, Australia, rests on a foundation of 180 steel tubes filled with reinforced concrete. Each tube is 150 feet long, three feet wide and can bear a load of 280 tons.

If a foundation is located in a fast-moving current or in strong tidal waters, it will probably be dug, or keyed, into the bedrock. San Francisco's Golden Gate Bridge, completed in 1937, is designed to withstand not only the powerful tides and pounding waves of the Pacific Ocean, but earthquake shocks as well. Its south pier, for example, stands on a base that has been dug 14.5 feet into bedrock. This keyed foundation prevents the structure from sliding out of place under extreme pressure. In regions where rivers run with pack ice, the profile of the underwater foundation is designed to offer the least obstructive face to the current. Here a prow-shaped cutwater may be used to protect the base and piers. For most foundations and piers, breakwaters of some sort provide protection from powerful currents, debris and scour.

Abutments for arched bridges tend to be quite simple in design. Usually they are heavy concrete footings angled away from the bridge to withstand the angular thrust of the arch. If possible, the abutments are built into solid rock that can bear the weight of the compressive arch. Where the ground is not solid enough, bridge builders sometimes inject a grout mixture of cement and water into the surrounding soil to stabilize the ground. Although simple, large arch-bridge abutments are usually massive; often they are as big as a multistory building.

Anchorages for the cables of suspension bridges are buried on each bank and must withstand a pulling force of thousands of tons. These anchorages are more than huge monoliths that counteract the tremendous pull of the cables through sheer weight; the tens of thousands of steel wires that comprise the cables are connected to steel bars embedded in the anchorages. An anchorage, if founded in solid bedrock, will work as a kind of diagonal plug. At the Forth Road Bridge, a suspension span in Scotland, for example, each anchorage is an almost 200-foot-long inclined tunnel that is bored into bedrock and filled with about 19,000 tons of concrete. The cables are attached to a steel slab that is installed at the face of each plug; the slab, in turn, is anchored back to the end of the plug with 114 prestressed steel cables. The anchorages for the Verrazano-Narrows Bridge, which are based in glacial sand, work on gravity—the result of their tremendous weight—as well as the friction between the concrete and the soil. Each anchorage is made up of a series of reinforced concrete blocks creating a single monolith that is 160 feet by 300 feet and 230 feet deep. About 130 feet of each anchorage is exposed above the ground; below the surface, each anchorage extends down to a level of about 100 feet.

Establishing the foundations of a bridge almost always involves large-scale construction in areas of water, and usually means excavating deep into a river, lake or ocean bed. A primary challenge is to keep the water at bay long enough to complete the building job. There are only a few practical methods known for working in such extreme conditions.

For construction carried on in rivers, the water's flow is diverted through tunnels or conduits. Once the river is redirected, the foundation site is dried and excavation to bedrock begins. Another, less costly method is to isolate the foundation site with a cofferdam. Here, interlocking steel sheets are driven into a riverbed to form

The 295-foot-long Salginatobel Bridge, near Schiers in the canton of Graubünden, Switzerland, is a classic example of a span made entirely of reinforced concrete. Designed by Swiss engineer Robert Maillart and completed in 1930, the bridge's deck is supported by a thin, hollow-box arch, contributing to its unique, streamlined form. The advantage of reinforced concrete is that thin slabs can replace the column and beams of steel bridges. The result is a lighter, cheaper, but still effective bridge.

a watertight, box-like enclosure. After the water trapped inside the cofferdam has been pumped out, the mud and loose rock is excavated down to the bedrock. The cofferdam is then filled with concrete, its walls acting as a form for the bridge foundation. Massive cofferdams were used to dig the foundations for both the Golden Gate and Verrazano-Narrows bridges.

Where piles make up the foundation, the common procedure is to drive hollow steel tubes to the necessary depth with giant pile drivers. The loose rock and earth within each hollow pile is scooped out mechanically and replaced with concrete. This, in effect, creates a solid compression column. In many cases, however, the ocean floor or riverbed above bedrock is too deep or soft to support a cofferdam, or too hard to allow the driving of piles. In these situations, caissons are used to excavate foundations. A caisson is a kind of elaborate, prefabricated cofferdam that is floated into position and then sunk. One type of caisson is made of a series of connected hollow tubes or cells; from above, the device looks something like a gigantic peg board. When the caisson reaches the ocean floor or riverbed, the sharp lower edges of the cells cut into the earth, trapping water and mud inside. The water is pumped out and the earth is removed with clamshell diggers; as the excavation proceeds, the caisson sinks further. Additional segments are added to the top of the caisson's walls to keep its upper section above water. The caisson continues to sink as the excavation proceeds, until bedrock is reached and sometimes penetrated. Then each of the cells—now long hollow tubes—is filled with concrete, creating a massive foundation.

Traditionally, men worked in the bottom of pneumatic caissons—huge digging chambers filled with compressed air to keep out the water and seeping muck during excavation. Unfortunately, the use of compressed air often led to caisson disease, also known as the bends, that caused the deaths of many 19th-Century workers. These conditions are a far cry from those of modern workers, who operate highly automated machinery from a control room situated at the top of the caisson; the only thing manual is the manipulation of the controls that activate the power shovels below.

BUILDING BEAM BRIDGES

Steel and reinforced concrete are now used interchangeably for the piers and decks of beam bridges—the simplest spanning structures to build. Beam bridges fit the bill where the gaps to be spanned are relatively short, or where builders can place a series of piers across the terrain and then connect multiple spans. Such beam bridges can be enormously long structures. One, in Hotseh, China, is 90 miles long; in the United States, the 24-mile-long Lake Pontchartrain automobile bridge crosses the Louisiana lake of the same name by connecting 2,170 short spans. Many sections of common elevated highways and expressways are in fact a series of connected beam bridges.

The construction of a simple beam or multiple-span beam bridge is as straightforward as its design. When the concrete and steel components are prefabricated, it is often merely a matter of using cranes to hoist the parts of the deck into position on their piers and then bolting them together. Where concrete for the bridge is made on site, scaffolding must be erected to support the structure temporarily. Forms are built atop the scaffolding and the concrete poured into them.

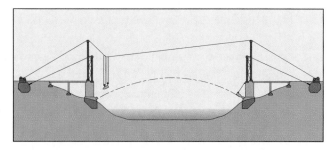

1. A cableway for an overhead crane is strung between temporary towers built atop the permanent bridge abutments. The towers are stayed back to temporary anchorages.

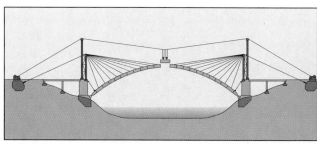

2. The crane installs steel segments of the arch symmetrically from both banks. Each segment is stayed with cables to the towers until the arch is connected at its crown.

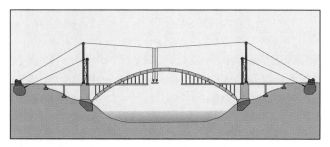

3. When the arch is complete, it becomes self-supporting and the temporary cables can safely be removed.

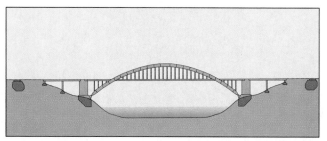

4. The bridge's deck is installed, after which the temporary towers are dismantled and removed.

Construction of a typical cantilever beam bridge is more complicated, since each span typically has three sections. The two main beams, anchored onshore, are supported somewhere along their length—usually near the middle—by a pier and extend toward the middle of the gap. A third, shorter girder joins the two arms. If the cantilever arms are the same length on each side of the supporting pier, the beam is in balance. On the other hand, if the pier is not in the middle of the beam, the downward push created by the longer arm of the span creates a countervailing upward push at the shore end, with the pier acting as the fulcrum. The massive concrete anchorages onshore counteract this upward force. Though the supporting piers are in compression, the main beams endure two stresses. They bend not by sagging in the middle, but at the ends; this puts the top of the beam in tension and the bottom in compression. To minimize the sagging, the beams may be thickened and thereby stiffened by a system of steel trusses along the deck that connects the ends of the beams to anchorages or piers.

One advantage of the cantilever bridge is that the piers can be spaced farther apart than in a standard beam bridge. This is important if the bridge crosses busy waterways where traffic cannot be impeded by closely spaced piers. Marine traffic was an important consideration in the design and construction of the two 1,700-foot spans of the Forth Railway Bridge in Scotland, then the world's longest cantilever bridge, which was completed in 1890. Only 27 years later, the single 1,800-foot span of the Quebec Bridge over the busy St. Lawrence River in Canada, surpassed its Scottish cousin and still holds the title as the longest single-cantilever span in the world. Both of these bridges are, like most cantilever spans, steel structures. Traditionally, concrete beams have not been strong enough in tension to

STAYING AN ARCH

Traditionally, arch bridges were built atop supporting scaffolding, or falsework, erected from the water level. But when water traffic cannot be obstructed by such scaffolding, other means are necessary. One solution for steel arch bridges is to use a cantilevered construction method, where the structure is temporarily supported, or stayed, from the shore by cables. The technique exemplifies the inherent strengths and weaknesses of the arch. The incomplete arch is unstable and can collapse without the support of falsework or stays. But when the two sides of the arch are joined, the structure becomes fully compressive and very strong. The diagrams at left illustrate the building of an arch using the cantilever method.

serve in cantilever bridges; only recently have designers begun to employ reinforced and prestressed concrete in a few modern spans, such as the Ganter Bridge in Switzerland, completed in 1980.

Many simple beam and cantilever bridges are distinguished by the extensive trussing along their decks. Not surprisingly, such spans are often called truss bridges. The familiar, triangular configurations of trusses give these bridges rigidity and strength, and make them ideal for withstanding particularly heavy live loads, such as those exerted by railroad traffic. While many train bridges today are truss spans, trussing is not a feature that is exclusive to railroad bridges. Indeed, trusses may appear on any type of bridge whenever the structure is in need of stiffening, including the decks of suspension bridges and arch bridges that will carry extremely heavy loads.

Whatever the materials that are being used, however, the construction of a cantilever bridge is necessarily a more delicate process than the building of a simple beam bridge. The nature of the two halves of the span atop their respective piers dictates the use of one of two construction techniques. The arms of the cantilever bridge can be built out from their respective shores, where they can be firmly supported in an abutment. Or, construction can begin atop the piers and extend evenly in each direction—toward the shore and toward the middle of the gap. As the spanning arm is extended toward the middle of the gap, it is supported by a developing lattice of steel trusswork. A cantilever bridge that is under construction looks extremely vulnerable as its arms reach out into space like seemingly unsupported wings. In fact, however, the trussing that makes the bridge extremely strong and rigid when it is complete also assures its integrity during its construction. Finally, the middle beam is raised into place from a barge by cranes mounted on the nearly completed bridge.

THE ENDURING ARCH

Stability and rigidity have always been a feature of arch bridges. However, the length of today's structures far surpasses traditional arch spans. The New River Gorge Bridge in West Virginia, for example, is a steel arch that spans almost 1,770 feet. And modern arch bridges are powerful structures. Australia's 1,650-foot-long Sydney Harbour Bridge carries four streetcar tracks, a roadway and two sidewalks for pedestrians—an ambitious undertaking for any bridge. The 945-foot-long Hell Gate Bridge in New York, another arch structure, needs even more strength to support the heavier load of its four railway tracks. But it is steel that underpins these strong and long-spanning arches; concrete has yet to match these distances. Among the longest reinforced concrete spans is the 1,267-foot-long Krk Bridge, for automobile traffic, near Zagreb in Yugoslavia.

But while steel spawned the development of long-spanning arches, reinforced and prestressed concretes provided the materials for an assortment of graceful new arch bridges that are valued almost as much for their artistic merit as their structural integrity. The Swiss bridge designer Robert Maillart was a well-known 20th-Century pioneer of reinforced concrete. His 295-foot-long Salginatobel Bridge over the Salgina Creek in Switzerland, completed in 1930, is still remarkable for its simple but elegant form. It is unencumbered by many vertical supports and by the trussing so familiar in steel structures.

Trestles, Tracks and Thrills

The roller coaster car screeches around tight corners and hurtles down drops that seem vertical at speeds approaching 65 miles per hour, its riders screaming, laughing or clutching the safety bar in terror.

Not many people are likely to experience such extreme emotions crossing a bridge, yet all types of roller coasters are essentially bridges. None resembles actual bridges more than the old-fashioned wooden roller coaster, popularized in 1927 by the Cyclone at Coney Island, New York, and now a national landmark. These structures are built in a similar style to the American wooden railroad trestles that span deep chasms with a horizontal and vertical framework strengthened by diagonal crossmembers. And although coaster tracks may look and feel rickety, the trestle gives the structure the strength inherent in each of its component triangles. No matter how much pushing or pulling force the cars exert on them, the triangular trusses are not deformed. However, while a bridge is designed to be nearly horizontal, roller coasters often have 60-degree inclines.

With riders fastened in, the cars are pulled up the incline track—that first big hill, usually more than 100 feet high—by a motor-driven chain hooked onto the cars' axles. The familiar clacking noise is a ratchet device, a safety mechanism that will grab the cars and stop them from rolling back if the chain should break. Once over the crown of this lift-hill, the downward force of gravity comes into play; the first drop gives the cars enough momentum to career around the rest of the track without further mechanical boost. The cars are in continuous motion, their velocity decreased when necessary by upward inclines and turns.

Although they are designed to give a sense of danger, in fact, roller coasters are very safe; they are becoming more so with computerized controls, and regular structural inspections using X rays and other sophisticated methods to pinpoint any incipient weakness.

With the application of brakes, the cars glide to a halt, lap bars are released and riders tumble out, exhilarated from their thrills on a bridge to nowhere.

Cars on the Great American Scream Machine, a wooden roller coaster at the Six Flags Over Georgia amusement park in Atlanta, zip along the 3,800-foot long track at speeds of up to 57 miles per hour.

The construction of arch bridges has traditionally relied heavily on two tools: falsework and cable stays. For many centuries, bridge builders always erected arch bridges in much the same manner: They carefully built the graceful form atop a supporting structure known as falsework, or formwork. The falsework held the sections of a developing arch in place until the uppermost block, the keystone, had finally been laid. Until the keystone had been placed, each side of the arch was inherently weak and vulnerable to collapse; with the keystone in place and the scaffolding removed, the arch became a powerful compressive structure. Today, however, this time-tested construction technique is often impractical for the construction of an arch bridge that will span a busy harbor or seaway where the traffic below the site cannot be interrupted; instead, a cantilevered method of building is now sometimes used.

It was James Buchanan Eads, builder of the multiple-arch steel bridge over the Mississippi, who first applied the idea of temporarily supporting an arch with stays during the construction of large bridges. With his system, a series of cables was used to connect the two arms of the arch to temporary supports on the shore or

At left are the massive cables of the 3,019-foot-long Akashi-Kaikyo suspension bridge in Japan under construction. The cables, from which the deck of the bridge will hang, are made of tens of thousands of wires the thickness of clothesline strung back and forth between opposite shores and over saddles atop the bridge's towers. The wires are compressed in parallel into three-foot-thick cables that are then banded. Laid end to end, the wires of a modern suspension bridge's cables would wrap around the Earth more than seven times.

to piers in midstream. This eliminated the need for falsework, since the bridge was supported from above rather than from below.

Though each of Eads' arches spanned only about 500 feet, his system gave other engineers the means to vault much wider gaps. Australia's busiest seaport was crossed in 1932 by the huge Sydney Harbour Bridge. Unlike the triple-span Eads Bridge, the Sydney Harbour Bridge is a single arch, spanning 1,650 feet; its deck is suspended by vertical struts from an overhead arch. Each arm of the arch was stayed while it was under construction, with 128 cables independently attached to temporary anchorages located behind the onshore abutments. The cables were strung through a U-shaped tunnel that had been dug almost 100 feet into solid rock behind each abutment. Close to 60 miles of cable was used for the job. Trolley-mounted creeper cranes worked their way up the stayed halves of the arch, adding more steel in front of them as they went. Once the half-arches were within inches of each other, the cables were loosened and the two sides moved together until they joined. The cranes moved back, erecting the vertical struts and the bridge's deck as they slid down the hump of the completed arch. Like all arches, the Sydney Harbour Bridge is very rigid and strong; upon its completion, that strength was demonstrated with considerable fanfare when the deck was loaded with 72 locomotives totaling 15 million pounds.

Unfortunately, the cantilever construction method does not work well for concrete bridges because of the material's low tensile strength. In many cases, falsework is still used for concrete structures, such as the 1,000-foot-long Gladesville Bridge spanning the western part of Sydney Harbour, where the four sections of its arch were precast and built atop a steel falsework. A gap was always left in this scaffolding to allow harbor traffic to pass through. Where scaffolding is impractical, other construction variations are used. One technique is to erect an arch formwork supported by cable stays and then to pour the concrete for the arch in layers. Each layer provides increased support for the next as the arch becomes stronger. On the other hand, bridges made of prestressed concrete, which is strong in tension, can be built using a cantilever system. Swiss bridge builder Christian Menn has used a cantilever construction technique for both prestressed concrete arch and cantilever beam bridges.

THE LONGEST BRIDGES IN THE WORLD
Nine years after the Eads Bridge opened for traffic in 1874, John Roebling's 1,595-foot-long Brooklyn Bridge was completed. A suspension span, and the longest bridge in the world at the time, Roebling's masterpiece featured a revolutionary use of steel: The overhead suspension cables were made of compressed strands of galvanized, high-strength steel wire, bundled together into a cable at the bridge site. The construction of the all-important overhead cables, although modernized, remains basically unchanged since Roebling first patented the process in the mid-1800s. Even before the Brooklyn Bridge, suspension spans had been the longest bridges in the world (except for brief reigns by two cantilever spans—the Firth of Forth bridge and the Quebec Bridge). Now, the longest suspension bridges are almost 4,000 feet long; in Japan the Akashi-Kaikyo suspension bridge, currently under construction and scheduled for completion in 1998, will soar more than a mile—a dizzying 6,529 feet.

The lofty towers of a suspension bridge, over which the cables will be draped, may be made of cast-in-place reinforced concrete or, more commonly, thousands of smaller high-strength steel cells—high, narrow boxes that stack like building blocks. Each tower sits on a reinforced concrete pedestal that is constructed atop the foundation. This pedestal, which in some instances reaches as high as 30 or 40 feet, keeps the legs of the tower from direct contact with the water. The legs of the towers are secured to the pedestal with bolts that can be up to three inches in diameter and more than 20 feet in length. The 70-story-high towers of the Verrazano-Narrows Bridge are constructed of some 10,000 separate steel cells—each about three-and-a-half feet square and up to 16 feet high—fastened together by more than a million bolts and three million rivets. Access to the towers is provided from the inside by 16 miles of ladders. To keep from getting lost within this labyrinth, those who built the towers wore miners' helmets with lights attached to the front and carried maps that could help them find their way to an exit.

When completed, the tower of a suspension bridge works as a single monolithic unit and looks something like a massive, elongated H-shape that has been joined on the top. Two horizontal struts connect the legs of the tower, one just below the level of the deck, the other at the top. A huge, curving saddle sits on the tower top to cradle the overhead cables as they pass over it. Between the horizontal struts, thick diagonal trusses are often used to add rigidity to the tower.

With ship traffic temporarily halted in the waterway below the site, wire ropes are laid by boat from one anchorage to the other, then hoisted into place by cranes mounted on the top of each newly constructed tower. Narrow catwalks, only a few feet wide and with walls of chain-link fencing, are hung from the newly suspended wire ropes to provide a wobbly footbridge for construction crews who will spend the next several months erecting the bridge's support cables. Eventually, more temporary cables, which will be used in the construction of the main cables, are hoisted atop the towers.

Although the process of making the main cables is called "spinning," no spinning actually takes place. Instead, each cable is built of tens of thousands of parallel wires put into place—wire by wire—using pulley-like carriages called traveling wheels. First, huge spools are brought to the site, each holding miles of high-tension wire. The end of the wire is attached to the end of a massive eyebar—a long, flat, steel bar that has been embedded deep within the concrete anchorage block on the shore. Then, powered by an electric motor, the traveling wheel carries the wire with it as it rides along the overhead temporary cables from the anchorage, over the towers, to the opposite bank where the wire is looped around an eyebar in the anchorage on that shore. In a single anchorage block there may be as many as 2,000 such eyebars, varying in length from 20 to 50 feet and weighing up to five tons apiece. The wheel travels back and forth from bank to bank carrying loop after loop of unbroken wire from the giant spools. Usually, there are several spools at each anchorage and four to six traveling wheels working simultaneously from each shore. When three or four hundred wires have been looped, they are compressed and banded together into a thicker bundle called a strand. More than 60 strands may then be hydraulically compressed and banded with huge collars to make up the three-foot-thick cable.

Workers stationed along the catwalks monitor the entire process, ensuring that each wire is properly positioned against those laid on previous passes of the wheel and that it hangs with the appropriate amount of sag, or give, between the towers. Each massive cable rides in the huge saddle positioned atop each tower. The cables are then coated with a corrosion-resistant material, wrapped in a flexible wire sheath for further protection against the weather and topped off with several coats of paint.

Once the cables are in place, the vertical suspenders from which the deck will eventually hang are attached to the bands that hold the wires of the cable together. With its cables and suspenders in place, the bridge is now ready to receive its deck. Formerly, this was done by building the deck out from each end or by hoisting the individual sections into place by overhead crane. Today, however, it is easier to assemble

The 1,740-foot Sunshine Skyway crossing Tampa Bay in Florida is an example of a cable-stayed span. In this type of bridge, the deck is connected directly to towers by means of high-tension steel cables and is supported, like a beam bridge, by onshore abutments at each end.

large sections, or blocks, of the deck on the shore and float each section on large barges into position beneath the main cables. With the help of cranes that are mounted on the bridge's main cables or on the barges, each section is then lifted into place and attached to its suspenders before being joined with high-strength bolts to the adjacent deck section.

Although the cables and suspenders of a suspension bridge are intended to be flexible and can sway sideways without causing damage to the bridge, the deck must be stiffened to protect it from the kind of wind-induced oscillations and torsion that destroyed the Tacoma Narrows Bridge. Since that 1940 calamity, engineers have paid special attention to the problem of deck flexibility. For some years engineers believed that the sheer weight of a suspension bridge would produce deck stiffness, an idea reinforced by the success of the tremendously heavy George Washington Bridge in New York. But engineers decided that the bridge had performed so well largely because it had been lucky enough not to be adversely affected by winds. Eventually, the George Washington Bridge was stiffened with trussing and the installation of a second deck. For less enormous suspension bridges, open trusswork offered the most support and the least wind resistance. However, the opening of England's Severn Bridge in 1966 marked the introduction of a new deck type designed especially to handle wind: the aerodynamic deck.

Here, plates of steel are joined together to create a hollow box section. The sides of each section are gradually tapered. When the box girders are joined together, the tapered sides form the outer edges of the bridge deck and present a narrower aerodynamic profile. Winds, therefore, pass above and below the deck with little resistance. Moreover, box girder decks require less steel than traditional trussed decks, reducing the overall weight and lightening the resulting load stress on the rest of the structure. As a result, the towers of a bridge built of box girders do not have to be as strong, its anchorages do not have to be as massive and its cables do not have to be as thick, making the bridge itself lighter and less expensive. This economy of design is evident in every aspect of both the Severn and the Humber bridges in England, with decks slung beneath cables that are thinner than those of older suspension bridges and towers that are pieced together from hollow blocks of reinforced concrete rather than steel cells.

Nevertheless, suspension bridges manage to consume enormous amounts of building materials. The construction of the Golden Gate Bridge, said to be a "battle of man against the sea," used enough steel to fill a freight train 20 miles long, enough concrete to pour a 25-foot-square block two miles high and enough wire to encircle the Earth three times. By its completion in 1937, the bridge weighed 100 million pounds.

One way of reducing the vast amounts of materials and money required to build long spans is to construct a variation of the suspension bridge, called the cable-stayed bridge. While not as long as their suspension cousins, cable-stayed bridges, often referred to as "medium-length spans," currently span 1,500 feet; and there are a number of 3,000-foot bridges under construction. The deck of the cable-stayed bridge hangs from two large towers, as in a suspension structure. However, unlike a suspension bridge, the deck is attached directly to the towers by a series of diagonal cables—which gives the bridge its name. This bypasses the need for the vertical suspenders and overhead cables that are anchored to huge blocks on the shores. Instead, the deck is supported partially by the diagonal cables, partially by the towers that also act as piers and partially by onshore abutments. The combination of these support mechanisms results in a stiffer span than the conventional suspension bridge.

MAINTAINING THE BRIDGE

Bridges require constant maintenance throughout their lifetime; it is even possible that substantial structural alterations and renovations will be made. Most parts

A 6,000-ton, prefabricated segment of deck for the Minami Bisan-Seto Bridge, a Japanese suspension span, is installed using two floating cranes. The cranes are each capable of lifting more than 3,000 tons. This method of prefabrication and deck-segment placement reduces construction time considerably. Traditionally, suspension bridge decks were laboriously constructed on site.

of a bridge require some sort of barrier against corrosion. Traditionally, paint has formed that front line of defense, and on steel bridges painting is a never-ending chore. Scotland's Forth Bridge, for example, has a full-time team of painters who apply 17 tons of paint each year in an ongoing maintenance program. Steel is often painted with zinc to shield it from corrosion, and in addition to zinc paint, there are various protective membranes that can be sprayed onto a bridge's superstructure. These dry to form a tough film that is not only waterproof, but also impervious to road deicing salt.

Despite such maintenance precautions, however, some of the world's cable-stayed bridges are in danger as a result of stress fatigue. The cables of Venezuela's Maracaibo Bridge, built in 1962, have already been replaced once and are due for a second refitting. Likewise, Germany's Köhlbrand Bridge, completed in 1975, needed a new set of cables after only three years; the problem was stress fatigue. The degradation of the beautiful Köhlbrand Bridge's cables gave further

impetus to the search to find better methods of protecting cables from the problem. Another persistent problem with the bridges is corrosion of the cables. Some promising new techniques for minimizing corrosion are under development, including plating the cables' component wires with zinc, encasing the cables in polyethylene pipes and surrounding the cables with polyethylene grout. Engineers also are experimenting with cables that are made of corrosion-proof graphite compounds instead of standard steel.

Such deterioration reinforces the need for regular maintenance and inspection of all bridges to pinpoint problems before they become serious. A more rigorous inspection program, for example, might have detected the hairline crack in an eyebar that was embedded in a concrete anchorage block, which led to the sudden failure in 1967 of the Silver Bridge. This suspension span in Point Pleasant, Ohio, plunged into the Ohio River, taking 46 motorists to their deaths. Today, bridge inspectors use portable ultrasound equipment to zero in on structural defects, and trained scuba divers carry out periodic inspections of bridge foundations in order to spot any deterioriation.

After 25 years of planning and nine years of construction, the Kojima-Sakaide bridge between the Japanese islands of Honshu and Shikoku, shown above, opened for business in 1988. More than eight miles long, the route features three suspension bridges, two cable-stayed bridges, a trussed cantilever bridge, three simple beam viaducts and a tunnel. Two more crossings are planned between Honshu and Shikoku, one of which will contain the longest suspension bridge in the world—a span over 6,500 feet long.

More advanced, computer-based monitoring systems are already in place on both the Golden Gate Bridge and Florida's cable-stayed Sunshine Skyway. The Golden Gate Bridge is equipped with a battery of tiltmeters linked to centralized computers. Installed in the wake of the devastating 1989 Loma Prieta earthquake, these are designed to detect movement of the bridge owing to scour, settling or earthquake tremor. The Sunshine Skyway's system is even more sophisticated, with more than 500 gauges and sensors, some embedded in the deck and piers, generating a stream of data designed to give bridge officials an early warning of trouble caused by wind, temperature, settling, impact or structural overload.

The day-to-day maintainence of a bridge keeps it operating safely and efficiently for the many decades of its life. But it does not generate the excitement that arises when plans for a new span are announced. Nor does it match the drama of cable spinning, or the raising of massive deck segments. Yet it is design, construction and maintenance that, in the end, combine to make a bridge work. As one official stated at the opening ceremony of the Golden Gate Bridge, "This bridge needs neither praise, eulogy, nor encomium. It speaks for itself."

DAMS

Years of labor make the moment possible. Every phase of planning and construction for the massive dam has been carried out on a colossal scale, involving hundreds of engineers, thousands of workers and tens of millions of dollars in contracts. Rivers have been diverted. Mountains have been sculpted. Instant towns have mushroomed, linked by roads, airports, even railways whose abrupt appearance in remote, often desolate regions might be dismissed first as optical illusions. Now the end, and the beginning, are at hand. Concrete for the last of thousands of massive blocks, appropriately known as monoliths, is about to be poured. The event, called a closure, is the occasion of a public ceremony. Though the dam's reservoir has been filling behind the wall for several years, a closure still must send a tingle along the spines of the crew whose blood and sweat have raised a hulking mountain out of blueprints and calculations. They have worked on the project for most of a decade; soon, the last of the formwork will be removed, and the mountain of concrete will either hold or not.

The dam will hold. These behemoths are not only the largest man-made structures in the world, they are also—by necessity—among the most successful. More than any other structure, dams are in direct confrontation with nature as they withstand the pressure exerted by trillions of gallons of water in a reservoir. The consequences of dam failure in the face of such force can be, and have been, catastrophic. The devastating results are on a scale more often associated with natural disasters than with a human engineering error. When a severe rainstorm caused excess reservoir water to wash over the crest of a dam in Pennsylvania in 1889, a wall of water inundated the nearby city of Johnstown, killing 2,000 people. The dam failed because its designers had not properly accounted for the possibility

Part of Canada's enormous James Bay hydroelectric project, La Grande 3, a massive hydroelectric dam on La Grande Rivière in Quebec, holds back trillions of gallons of water in a huge reservoir. Some of that stored water is channeled through the dam into a powerhouse where it generates hydroelectric power. Dams throughout the world use their stored water to supply abundant and potable drinking water, to assure reliable agricultural irrigation and to provide energy for industry.

ANATOMY OF A DAM

The illustration below shows one type of dam—a concrete gravity dam that uses its massive weight to resist the powerful push of the water in its reservoir. To do so, the dam is firmly anchored to a stable foundation. A vertical wall, shown in cross section, holds back the water, while spillways control the release of water from the reservoir. Giant penstocks direct falling water through the interior of the dam towards a hydroelectric generating plant at its base.

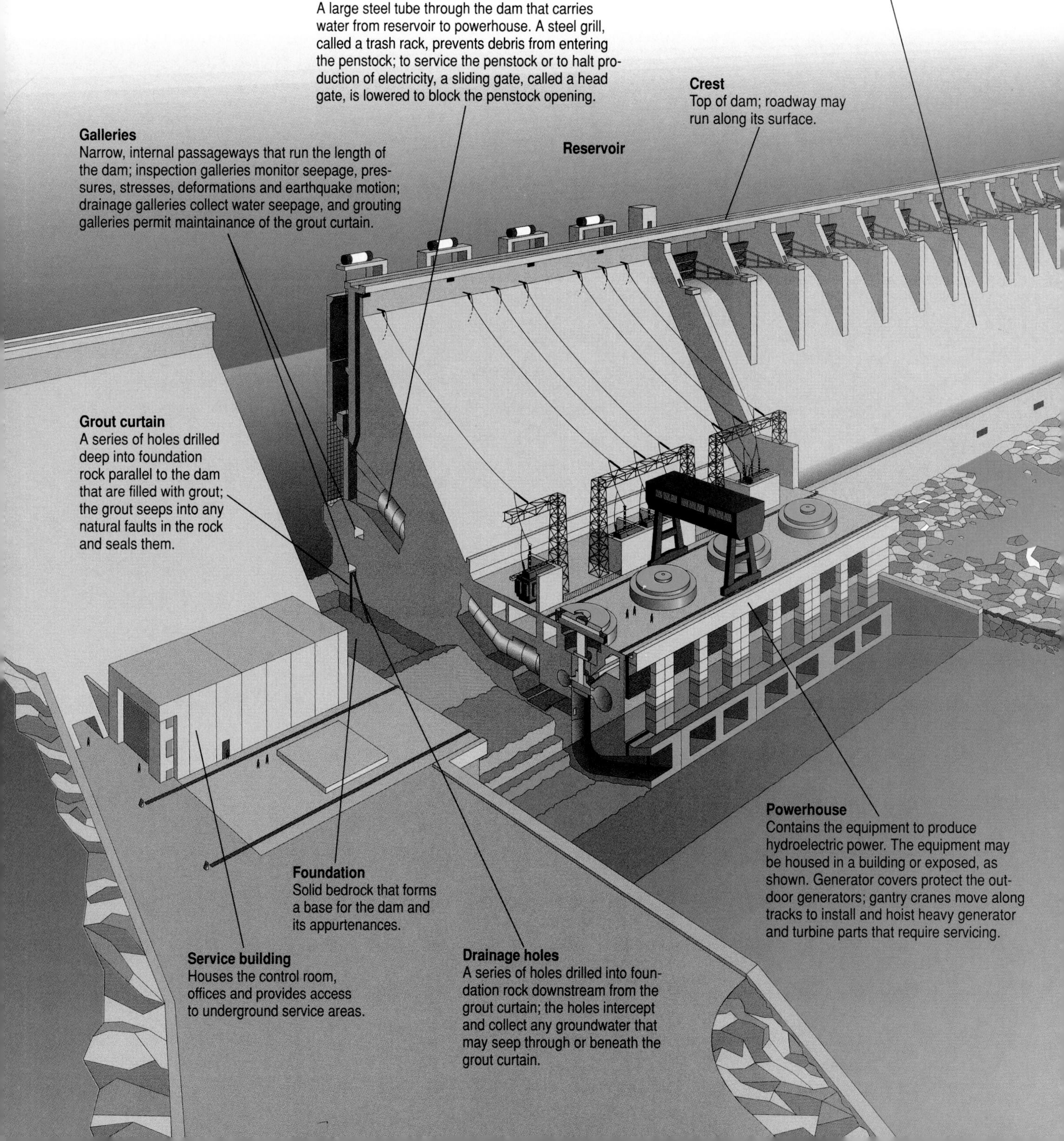

Spillway
A structure on the dam or to one side of it that releases excess water from the reservoir. To prevent erosion of the dam, a curved lip at the bottom of the spillway dissipates the energy of the water.

Penstock
A large steel tube through the dam that carries water from reservoir to powerhouse. A steel grill, called a trash rack, prevents debris from entering the penstock; to service the penstock or to halt production of electricity, a sliding gate, called a head gate, is lowered to block the penstock opening.

Crest
Top of dam; roadway may run along its surface.

Galleries
Narrow, internal passageways that run the length of the dam; inspection galleries monitor seepage, pressures, stresses, deformations and earthquake motion; drainage galleries collect water seepage, and grouting galleries permit maintainance of the grout curtain.

Reservoir

Grout curtain
A series of holes drilled deep into foundation rock parallel to the dam that are filled with grout; the grout seeps into any natural faults in the rock and seals them.

Powerhouse
Contains the equipment to produce hydroelectric power. The equipment may be housed in a building or exposed, as shown. Generator covers protect the outdoor generators; gantry cranes move along tracks to install and hoist heavy generator and turbine parts that require servicing.

Foundation
Solid bedrock that forms a base for the dam and its appurtenances.

Service building
Houses the control room, offices and provides access to underground service areas.

Drainage holes
A series of holes drilled into foundation rock downstream from the grout curtain; the holes intercept and collect any groundwater that may seep through or beneath the grout curtain.

of such a high level of flood waters; they underestimated the safety factor and did not build a high enough dam. To compound and prolong the catastrophe, the dam was made of earthen materials. As the torrent rushed over the top it eroded away the dam, which released more and more water.

Basic physics—Newton's third law—teaches that all structures must push back at the physical forces of nature with at least equal pressure. To stop the flow of a river then, a dam that matches the strength of that river must be built. And since required safety standards impose additional structural conditions as well, the dam must be even stronger than the power of the river.

But a dam is more than just a giant plug that stops a river from flowing, although it is a common enough belief. As one unimpressed old-timer who watched the construction of the Grand Coulee Dam on the Columbia River in Washington commented: "If the Almighty can dam up the Columbia with a cake of ice, man can do it with a piece of concrete." The old-timer's sense of the structure as a giant waterstop plug is sound as far as it goes, but does not explain the many complicated principles and combinations of forces that determine modern dam design. In terms of physics, a dam submits to and resists a variety of natural and man-made loads. Though different types of dams work on very specific principles, the basics are constant: Hydrostatic pressure—the force exerted by trillions of gallons of reservoir water pushing on the dam—is channeled through the main body of the dam into its rock foundation in the canyon or valley bottom and walls. This basic load bearing is key to understanding how dams work, but the completed structure also relies on a careful mix of foundation engineering, knowledge of materials behavior and specialized construction techniques, all of which allow dams to be bigger, safer and more productive.

THE LARGEST STRUCTURES ON EARTH

All dams fall into one of two general categories, depending on how they manage the precious resource with which they are inextricably linked. Diversion dams, the first category, are designed to divert or deflect water from a stream or river into a canal, conduit, treatment plant or small hydroelectric facility. These modest dams are not designed to block an entire waterway. During a flood, for example, water may wash over the dams without damage. Numerous, and often in or near cities, diversion dams are largely invisible on the landscape.

Storage dams—the second category—are hard to miss, despite the fact that most of their bulk and height is submerged. The highest of these dams exceed 1,000 feet, taller than most skyscrapers. The longest, though not as high, are sometimes more than 10 miles long. Storage dams retain water for long-term use. A reservoir, the artificial lake created behind the dam, can serve as a reliable source of drinking or irrigation water. Storage dams can prevent flooding during rainy seasons and impound seasonal run-off for use during droughts. And these dams also may perform the high-profile function of generating clean and renewable hydroelectricity.

Storage dams themselves fall into two basic categories: the massive and the structural. Massive dams work by virtue of their own enormous size or weight, often tens of millions of tons. They are also known as gravity dams because they rely on that most basic of forces to do the job. The largest and most common massive dam is the embankment dam, the workhorse of all dam designs. In cross sec-

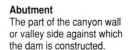

Abutment
The part of the canyon wall or valley side against which the dam is constructed.

tion it looks like a mountain, with a massive base, sloping sides and a peak. Indeed, it is made largely of natural materials such as clay, silt and sand, or broken rock, or a combination thereof. And like all mountains, the embankment dam's stability is grounded in its broad foundation, typically five times its height, which may range from a few dozen to more than 1,000 feet.

Usually, earth or rockfill constitutes the thick outer shell of the embankment dam. An upstream face of stone, or other materials that will resist erosion, shields the shell against the wear of water. In the center of the shell is a water resistant core, made of clay or other fine materials, that juts down into a wide trough, called a cut-off trench, excavated through the river bottom material to bedrock. Beneath the trench is a vertical man-made barrier, known as a grout curtain, that extends deep below the foundation of the dam, sometimes hundreds of feet into the bedrock. Almost all storage dams rely on these barriers to reduce the flow of water through the foundation and under the dam. Since all embankment dams do leak to some extent, most include an internal drainage system to siphon off the water that unavoidably seeps through the shell and core.

A concrete gravity dam is essentially a huge retaining wall. In profile, this dam looks like a right-angled triangle, with a vast vertical face meeting the reservoir's wall of water head on and a sloping side looking downstream. Originally made of masonry, these dams have become so huge that they are now concrete structures, since only concrete could provide the material strength to erect such a famed giant as the Grand Coulee Dam on the Columbia River in Washington state. Eight years of labor and nearly 12 million cubic yards of concrete—enough to pave a highway halfway across North America—went into the construction of the 550-foot-high, 4,173-foot-long Coulee dam, which was completed in 1942.

But concrete gravity dams are not all brawn. Such a structure is proportioned so that the main loads—its own weight and the push of the water in the reservoir are played off against each other. Hydrostatic pressure is horizontal, while the pull of gravity is vertical. Designers calculate the stress imposed by these forces, factoring in the weight of the dam, and intended wall height and area. Here, the same principle—the middle third rule—that applies to loads on columns and arches *(page 20)* is applicable again. The design must direct the load diagonally through the middle third of the thickness of the dam and into the foundation. If the load stays within this region, all the forces acting on the dam will stay in balance and in compression; if the load falls outside the middle third, the dam could undergo bending stress.

Engineers must also deal with the stress caused by uplift. Groundwater, the subterranean water that flows in streams or seeps into cracks and fissures in the rock, can make its way into the foundation of a dam. When the water is pressurized by the weight of the reservoir, it will force its way through the cracks to exert upward vertical pressure on the dam, upsetting the careful balancing act between hydrostatic pressure and gravity by degrading the foundation. This can lead to cracking and eventually may cause the dam to split; part of the dam building process, therefore, includes the construction of grout curtain barriers under the base and into the rock foundation below.

Though mammoth gravity dams are relatively simple in design, they also are extremely expensive to build, an impediment that hastened the search for new

THE BIG FOUR

Illustrated at right are the four basic dam types: embankment, concrete gravity, arch and buttress. Each is designed to withstand an assortment of powerful forces including uplift, water erosion and earthquake activity. Embankment and concrete gravity dams work primarily by virtue of their size and weight, and so are called massive dams. Arch and buttress dams rely on shape as well as size, and are known as structural dams. Of all the forces that dams must withstand, foremost is the horizontal hydrostatic force of the reservoir water against the upstream face of the dam *(black arrows)*. In the case of a buttress dam, the force of the water also has a small vertical component which pushes down on the sloping, waterproof upstream face. The red arrows represent the dam's resistance to sliding downstream. This is provided by the friction between the base of the dam and the foundation and by any notch, or key, dug into the foundation. The purple arrows show the way the dam's design and weight withstand the hydrostatic pressure. The blue arrows show the counteraction of the foundation to the applied forces.

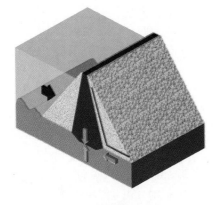

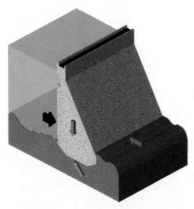

Embankment dam
Massive embankment dams are made of natural materials, either rock or earth, piled into a mountain-like form with sloping sides. As the reservoir submerges the upstream face of the dam, water seeps into the voids between the rock and earth particles, giving them a buoyancy or small uplift. After seeping through the core, the water is channeled away by the internal drainage system. The structure is stabilized by its extremely wide base, which creates a large area of friction against the foundation.

Concrete gravity dam
Most concrete gravity dams expose a nearly vertical wall to the push of the reservoir, absorbing the load straight on. The combination of the weight, shape and materials that make up the dam assure that hydrostatic force—the push of the reservoir—is channeled downward through the middle third of the dam's base. This load and the enormous weight of the structure is absorbed by a solid rock foundation.

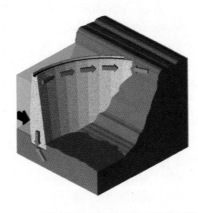

Arch dam
It is shape, rather than size or weight, that keeps a concrete arch dam upright. The hydrostatic pressure is channeled by the dam horizontally through its arching form, as seen in the half-arch shown. The resulting end thrust is met and absorbed by the canyon walls. Friction between the base of the dam and its foundation plays a minor role in resisting the downstream push of the water.

Buttress dam
The buttress dam is actually hollow, made up of an upstream sloping concrete wall that is supported by a series of triangular-shaped slabs of concrete, or buttresses, which are often separated by struts to keep them from moving toward each other. As well as a horizontal force, the water pressure from the reservoir applies a vertical force to the upstream face, helping to hold the dam in place. These forces, along with the weight of the dam, are transmitted through the middle third of each buttress into the foundation.

dam designs and more affordable construction methods. A solution is structural dams that, while heavy in their own right, work on different principles. Shape, rather than size, controls the forces exerted by the reservoir. Often two-thirds less construction materials are required for these dams, making construction a cheaper proposition, but the amount of engineering involved increases in proportion. The arch dam appears perilously tall and thin, but works like all arches: Compressive pressure applied to the crown of the arch is directed along its curving form and then redirected, as end thrust, outward into strong abutments. With this design, the hydrostatic pressure is directed along a horizontal plane rather than a vertical one and is transferred into the canyon walls. Sometimes, if the rock in these walls is not up to the task, artificial supports, called thrust blocks, are constructed to corral and dissipate the arch's pushing force.

A fascinating new development in arch dam technology during the 1960s was the evolution of the double-curvature shape that made it possible to build higher and thinner dams. Traditionally, arch dams curved from side to side, but not from top to bottom, leaving a vertical face exposed to the reservoir pressure. The double-curvature design curves the dam both horizontally and vertically, like a spoon.

This design distributes the loads more evenly across the canyon bottom and walls, reducing the stresses that normally would be present on a vertical wall.

Buttress dams, the other main structural design, evoke the secular spirit—not to mention the physics—of the Gothic cathedral. Massive buttresses on the downstream side of the dam brace the structure against the enormous horizontal push of the reservoir's water. But buttress dams also use the reservoir ingeniously to reduce horizontal push. A sloping upstream slab, or face, is pushed downward by vertical water pressure, effectively playing horizontal hydrostatic pressure and the vertical pull of gravity against each other. As with a concrete gravity dam, the hydrostatic load must be distributed down through the buttress to the foundation. However, the vertical load on the upstream slope reduces the need for as heavy a structure as a concrete gravity dam. Another variant of this design, the multiple-arch dam, features a series of connected arches supported by downstream buttresses formed and enlarged where the arches join. This can give the appearance of a section from a classic Roman aqueduct turned on its side. Used in wide valleys where a single arch might be too difficult to construct, they can be enormous structures. The 6-million-ton Daniel Johnson Dam on the Manicougan River in Quebec comprises 13 arches, the highest and largest of which is 703 feet tall.

When a dam stores water, it must have a means to release or redirect it for other uses. Each dam is designed to hold back a maximum amount of water in the reservoir, but requires a means for quickly releasing any excesses that result from flooding or rainstorms. Spillways handle this job, guiding the flow of excess water into the river system below the dam. Some spillways direct the excess water from the reservoir into tunnels bored into the canyon walls; others through conduits that pierce the base of the dam; and still others along surface chutes that wind around the site. All outlets are lined with smooth concrete to prevent water erosion since rushing water can be extremely abrasive. The tunnels flanking the Hoover Dam can bypass an amount of water equal to the volume that pours over Niagara Falls, and with twice the vertical drop.

Another kind of spillway is the overflow variety, a spectacular waterfall plunging right over the center of the wall. This design disposes of water quickly, and the effect is visually thrilling. The dam does, however, require extra reinforcement to ensure that the water churning the foundation causes no damage. Every second, a million gallons of water can pour through such a spillway, pounding remorselessly at the bottom. To dissipate this force, an upturned concrete lip at the base of the dam curls the water like a wave, directing its energy in an upward motion. The destructive force of the falling water can also be broken by baffle piers, giant concrete cubes across the downstream base that break and slow the flow.

WINDOWS ON A SUBTERRANEAN WORLD

Naturally, any dam has an environmental impact. Altering the flow of a river affects local watersheds and far-flung drainage systems. The creation of lake-sized reservoirs will flood territory populated by wildlife and often people. The benefits must always be weighed against the costs; sometimes the decisions are dubious. The reservoir of the Aswan High Dam in Egypt, for instance, not only forced the resettlement of 70,000 people, but also initiated a frantic, worldwide effort to salvage as many ancient artifacts as possible before the sites were doomed to underwater oblivion. Moreover, age-old flood rhythms were altered drastically by the dam, precipitating a crisis in Egyptian agriculture.

Several other very practical matters affect site selection: foundation conditions, time constraints and the accessibility of equipment and a skilled labor force. Availability of materials is an additional factor. Embankment dams can be cheaper than other types, but only if suitable local materials—earth and rock—are readily available. If the material is too far from the site, the cost of trucking it in becomes prohibitive. A moderately sized 40-million-cubic-yard earth fill dam would require hundreds of thousands of truckloads of fill. Concrete gravity dams are always more expensive, especially if the site is isolated. Millions of cubic yards of concrete must be mixed from trucked cement and local aggregates, poured, and allowed to set—a labor-intensive operation.

The physical nature of the terrain is obviously central to the choice of site because both topography and geology dictate what type of dam can be built. Arch designs are ideal for narrow V- or U-shaped gorges where strong rock walls can serve as abutments to absorb the strong end thrust. A valley underlain with consistent rock is essential to support heavy concrete gravity dams, while earthen foun-

Nature's Master Builders

In constructing their remarkable dams, beavers instinctively take into account the physical principles of hydrostatic pressure and structural weight. And like human engineers, they build their dams with foundations dug into the stream bed, upstream walls and water outlets to control the reservoir level. What is more, they are very quick about it: A family of industrious beavers can build a 30-foot-long dam in about a week. But that is only the beginning, for the beavers must work constantly, as in the photo at left, to patch and otherwise maintain their creation. Properly cared for, a dam will last for many generations. When a colony of beavers sets to work, the results can be impressive indeed—witness the magnificent edifice that stretches 2,100 feet across Montana's Jefferson River.

The diagram at left shows a cross section of the upstream side of a dam built by the Rhone Beaver, a European species. Beavers use their chisel-like incisor teeth to cut a variety of sticks and logs. The larger pieces—usually logs or even stumps—are inserted vertically into the river bed to establish a strong foundation. Depending on the nature of the river bottom, the foundation is often anchored with mud and rocks. Finer sticks are piled up against the upstream side of this foundation to provide the bulk of the dam.

dations, which could include clay, silts and gravel, might suffice for embankment dams. Often, a potential site will suit a number of types of dams and will give rise to a hybrid such as the Grandval Dam in France, which features an amalgam of buttress, arch and gravity features.

Once a potential site is located, the ground is thoroughly charted and evaluated by geologists. The first stage of site investigation is remote sensing, an examination of the site from the air. Traditionally, aerial photographs were taken; now images from earth-orbiting satellites are sometimes used. This topographical investigation may reveal large faults or other irregularities on the surface that are indicative of fault zones. Another process, infrared photography, detects slight differences in heat emissions from the Earth that may also expose smaller faults, fissures or fractures in the rock below the surface.

But topographical investigation inevitably takes a back seat to the less dramatic but more comprehensive ground testing. The dam builder's real windows into the subterranean world are hundreds of boreholes. Diamond-studded core bits, normally anywhere from 7/8ths to 2 1/8th inches in diameter, or larger for special purposes, drill hundreds of feet below the surface of the Earth. When the core samples are extracted, they provide a narrow profile of the geological formations below. They reveal exactly where bedrock lies, where underground streams flow and where large faults have split the rock. They also tell much about whether the density of various rock or soil layers will provide adequate resistance to the compressive and shear stresses the dam will exert, and how extensively water seeps through the rock. The same sort of investigations are made into the canyon walls, which must absorb immense pressures.

THE ALL-IMPORTANT COFFERDAM
With the site selected and mapped—both above and below ground—the slow process of building the huge dam begins. The entire project might take as long as a decade. The first task, no mean feat in itself, is to divert the river around the construction site. In order to do this, it is often necessary to construct one or more temporary structures, called cofferdams. Sometimes the cofferdam is a simple earth and gravel dike, which redirects the entire flow of a river into channels or tunnels that bypass the site, leaving it relatively dry and exposed enough for construction. A bypass tunnel is a major engineering effort, especially if it has to be driven through the solid rock of a narrow river gorge. The tunnel will need to handle not only the normal flow of the river but, in many cases, the added volume from any spring run-off as well. It is not unusual for the tunnels for major dam projects to be 40 or 50 feet in diameter in order to handle this flow.

Often, a cofferdam will be designed to enclose part of the construction area in a kind of watertight corral. Cofferdams, which sometimes look like huge barrels plunked down in midstream, are made of interlocking steel sheets driven into the river bed. The trapped water is pumped out and construction begins. As the dam wall is built inside the corral, new cofferdams are extended further across the river. Taming the mighty Columbia River—to start the Grand Coulee Dam—tested the ingenuity of engineers. Two U-shaped cofferdams, constructed of steel piling and timber, were built out from each bank. The water funneled through a 500-foot gap, which was wide enough to allow the river to flow. The vast areas inside the

The massive Hoover Dam, which controls the flow of the Colorado River, near Boulder City, Nevada, is still considered an engineering marvel nearly 60 years after its completion. Although it is heavy enough to function as a gravity structure, it also arches into solid rock canyon walls for support. When it was completed in 1936 the 726-foot Hoover was twice as high and—at 1,200 feet long and 660 feet thick at the base—twice as large as any previous concrete structure. The sheer bulk of the dam necessitated a host of new construction techniques, many of which are now commonplace.

Thames Tamer

L ike a set of giant, burnished stepping stones, seven steel-helmeted piers stretch across the Thames River at Woolwich Reach, eight miles east of London. At the feet of these solid sentinels are six curved steel plates lying flat in concrete sills flush with the river bottom. Within 30 minutes they can rise from the sludge to protect the city of London and its lowlying environs from flooding.

During a spring tide—the highest high tide, which occurs twice a month—a hump of water about eight feet high may originate in the Atlantic Ocean, funnel down into the North Sea and enter the Thames estuary. Then strong winds can blow the water 60 miles upriver to London. In 1953, London was spared the fury of such a surge tide at the expense of downriver areas, which were flooded when seawalls collapsed. Three hundred people died.

Although the Thames banks had been built up over the years to hold back floodwaters, the ultimate solution was a structure that could combine the protection of a permanent dam with the versatility of a drawbridge. In 1974 construction began on the Thames Barrier, a colossal movable barrier that in 1982 was ready to stop the surges.

When a flood appears imminent, river traffic is diverted and the gates unlocked and raised into position in succession. The biggest of the six gates are 200 feet wide, weigh 3,200 tons and can hold back 9,000 tons of water. When the two center gates, the last to rise, are in position, an 80-foot high barrier is formed. It stays in place as long as there is danger—usually from two to five hours. Then the gates are raised slightly off the sill to allow water to pass under the bottom of each gate. Once the water levels on both sides of the gates are almost level, the gates return to their invisible resting place.

The Thames Barrier, shown here in its raised position, spans the river from its north to south shore, forming an impenetrable barrier. The six gates seen between the 200-foot-long piers usually stay on the river bottom so that shipping can continue.

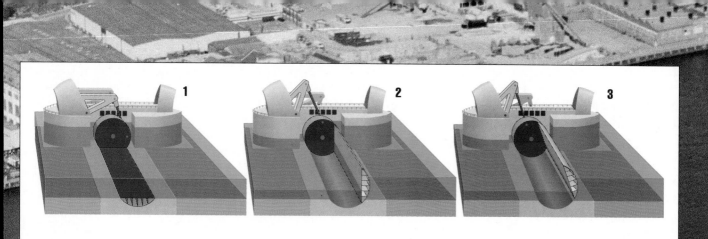

RAISING A GATE

The illustrations above show the gate-raising procedure. To allow ships to travel past the barrier, the gate normally lies flat on the river bottom (1), with its curved face fitting into the concrete sill. To raise the gate, the hydraulically operated machinery in two adjacent piers lifts rocker beams, which pull on arms attached to each end of the gate. The gate then rotates up into its vertical, raised position (2). After flood danger has passed, the gate is lifted three to four feet above the sill (3) to let water pass underneath, gradually returning the river flow to normalcy. Then the gate is lowered and returned to the river bottom. During maintenance, the gate can be pulled up until it is in a horizontal position above the water.

cofferdams were then pumped dry and work began, with the walls inching toward each other in the middle of the river. Eventually two more cofferdams above and below the site forced the river into new spillways, allowing the walls to be joined.

DOWN TO ROCK BOTTOM

With the river temporarily diverted and the construction site dried out, heavy labor begins. Soil and fragmented rock are removed from the riverbed, exposing the stable geological layer that will provide the foundation for the dam. In excavating for the Grand Coulee, 13 million cubic yards of glacial material had to be removed from the bed of the Columbia River. This undertaking required the construction of a 1.5-mile-long conveyor belt linking the dam site to a nearby canyon where the debris was deposited. At the site of the Daniel Johnson Dam in Quebec, workers had to remove 150 feet of alluvial deposits below the riverbed in order to reach the hard-rock bottom of the gorge.

Blasting, usually necessary during excavation, must be done with care to avoid splitting the rock deep below the surface. The blasted material is removed with heavy-duty graders, bulldozers and loaders. Most dams sit in trenches dug to, or into, the bedrock. For embankment dams these are the cutoff trenches. Alternatively, a less-expensive, narrower excavation, called a diaphragm, might be constructed beneath the dam. Under arch and concrete gravity dams the trenches are called keys, because in profile they exhibit the jagged notches seen on house keys. The increased surface area of the notches provides more friction, and therefore more stability, between the base of the dam and the foundation. With the vertically extended core of an embankment dam, or the central portion of a concrete dam, sitting in the trench, a barrier has been created to protect against reservoir seepage. The trench also provides a friction area, or notch, between structure and foundation that anchors the dam and prevents it from sliding downriver.

Canyon walls also are excavated and carefully cut into notches, called keyways. The walls, which act as abutments, must be sculpted so that the concrete blocks of a concrete dam and the core of an embankment dam fit snugly into the canyon. This is a hazardous job that requires crews to operate large crawler drills that work their way down the face of the gorge carving into the rock. Often, workers dangle in harnesses from ropes, chipping out weak spots and smoothing the rock. Throughout the treatment of the foundation and walls, a massive cleaning job is done. Workers armed with powerful hoses and brushes scour and clean the exposed rock so that eventually the dam will interlock neatly and unfailingly with the foundation. This is particularly important because the concrete or natural fill must bond almost seamlessly with the natural rock.

Before the first bucket of concrete is poured, or natural fill trucked in, dam builders go to great lengths to stabilize, or consolidate, the foundation. The dam's weight can be so much that the load may cause shear or excessive compression in the rock foundation. The consolidation process is achieved by a system called grouting. Grout is a fluid or semifluid mixture, most commonly water mixed with Portland cement, that is pumped under pressure into holes drilled into the foundation rock. The grout mixture will spread out, slowly filling joints, faults and cavities in the area. When it hardens the effect is further stabilization of the ground so that it will behave as a single foundation unit. It is a delicate process; too much pressure can fracture or shift some rock formations. The first stage of the process, called consolidation grouting, is usually done down to a level of 20 to 30 feet. For concrete gravity dams, the grouting area encompasses the entire base of the structure; for embankment dams grouting is done only below the central core, sealing all openings into which fine soil from the core might wash. In both cases, consolidation grouting creates a barrier to underground water.

Next, workers create an even deeper barrier with what is called curtain grouting. Often the depth of a grout curtain will equal the dam's height. Grouting does not entirely prevent groundwater from seeping under the foundation, but it reduces the amount and pressure significantly. When grouting is completed, the result should be a firm, consolidated foundation that unfailingly supports the structure.

In some cases a long, narrow passageway called a gallery is built into the dam near the foundation. Like the tunnels that lead deep within a pyramid, this gallery allows engineers into the heart of the dam from where they can perform inspections, carry out further grouting, or drill drainage holes throughout the life of the dam. In larger dams there may be several miles of inspection galleries.

EVERY PROJECT UNIQUE

Constructing a dam is a massive undertaking, underscored by gargantuan facts and figures. The Grand Coulee Dam used almost three times more material in its construction than the Great Pyramid at Cheops. Nine years in the making, this concrete gravity leviathan is 500 feet thick at the base, 550 feet high, and 4,173 feet long. Twelve million cubic yards of concrete were poured into the structure. Yet in terms of volume, most concrete gravity dams are dwarfed by embankment dams. The massive 754-foot-high Oroville Dam in California, for example, used a quarter of a million cubic yards of concrete just to create a huge anchor block for the clay gravel core. Although it does not even make the top 10 list of the world's largest embankment dams, it holds 78 million cubic yards of earth.

More monumental still is the Itaipu Dam in Brazil, an embankment-concrete structure that gives a sense of the vast numbers of people, and incredible expense involved in the construction of the largest dams. This hydroelectric structure, built across the Parana River between Brazil and Paraguay, is 643 feet high and 4.5 miles long. During the nearly nine years of its construction the epic enterprise employed a total of 29,000 people; skilled laborers who toiled day and night. The site had 9,500 on-site housing units, educational facilities for the workers' children and canteens that could dish up nearly a million meals a month. Almost 16 million cubic yards of concrete were sunk into Itaipu, more than five times the volume of Nevada's giant concrete arched-gravity Hoover Dam, the highest dam in the world when it was completed in 1936. Construction began on the Itaipu in 1975. Though it was structurally complete by 1984, is was another five years before it was fully operational. Its total cost was a staggering 13 billion dollars.

Whatever the statistics, each dam is a unique project, built to its own particular specifications. Nevertheless, there are some common construction techniques. Earthen embankment dams are built up incrementally in layers, typically using a rolled-fill method. Here, layer upon layer of earth is trucked in, dumped and compressed by large machines called compactors. Giant drums fitted with protruding knobs roll back and forth across the material until it is fully compressed and water resistant. The less common rockfill dams may also include a concrete or asphalt face to shun water on the upstream side.

Concrete gravity structures, on the other hand, are constructed with a kind of building-block approach using huge concrete monoliths. Massive forms are erected, then slowly filled with concrete. Each form, a small construction in itself, is bolted together and held in place by a powerful steel frame reinforced with vertical beams. The schedule for pouring concrete is tight and crews work frantically to finish the forms before the first of many eight-cubic-yard buckets of concrete descend from a suspended cableway above.

When about 20 inches of concrete has been poured, a heavy-duty vibrator is immersed in the mix inside the form. The shaking and vibrating helps settle and compact the concrete evenly before the next layer is added. When the forms are removed, a monolith—almost 8 feet high and 50 feet by 50 feet in area—is exposed. Within a few days the crew must quickly construct another form and blast away the thin top layer of the previous monolith with a water hose—a process called green-cutting—to expose the uncured concrete underneath. Then, when the next bucket of new concrete arrives and is poured, it bonds seamlessly with the block below. Many of these blocks are poured each day. The dam is swarming day and night with carpenters, welders, plumbers and pipefitters working on the formwork for the next day's pours.

The process is repeated thousands of times to create the bulk of the dam. When the concrete has fully set, the vertical joints between the huge monoliths are filled with grout, which consolidates them into a unified structure that acts as a whole under temperature changes and hydrostatic pressure. To ensure that the blocks are watertight, copper strips and flexible plastic seals called waterstops are embedded across the joints between the monoliths near the outside of the upstream face.

THE NEW AGE OF CONCRETE

If the gravity dam is constructed like a kind of giant building-block set, the arch dam is put together more like a jigsaw puzzle. Converting the smoothly rendered lines on a blueprint into a gently curving, solid concrete arch is a marvel of geometric precision. An arch dam diminishes in thickness as it heightens. Traditionally, thousands of individual molds were built for the concrete blocks of an arch dam. Today's steel forms can be reconfigured quickly, permitting the customization essential for creating the curving shape of the wall and the glove-tight fit between the concrete and canyon walls. Special joints and water-stops also are required.

Truly huge concrete dams became commonplace after the triumph of the Hoover Dam project in the early 1930s. The size of the Hoover revolutionized many aspects of construction, but none more so than the pouring, mixing and cooling of concrete. For example, since concrete generates heat as it hardens, the temperature of the material as it sets is a constant concern. Unless the process is carefully controlled, heat actually can remain in the material for years, causing a block of concrete to expand and then contract when it finally sets. Engineers at the Hoover site hit on the idea of embedding pipes into each block during construction so that they could reduce the temperature of the concrete by circulating cool water through the pipes. This is now a common part of dam construction. Since the Hoover project, the mixing ratios of various ingredients, techniques for vibrating

Seen under construction here are the concrete spillway and powerhouse of the Barkley Dam, a multipurpose embankment structure on the Cumberland River in Kentucky. It provides hydroelectricity, flood control, recreation and has a lock for boats navigating the river. The dam is a combination earthfill and concrete gravity type and is almost two miles long. Construction for most large dams goes on nonstop, day and night; huge floodlights illuminate the entire site, which seems to dwarf the army of workers and their equipment. Huge concrete blocks, or monoliths, are used to build the spillway, at right, and for the concrete section of the dam.

and pouring at different speeds and the understanding of the chemical properties of concrete have been improved continuously. Sophisticated computerized mixers are used to measure precise quantities of cement, aggregate and water. Refrigeration plants and coolants keep the mixed concrete cool; sometimes ice chips are used in the mix instead of water.

But concrete dams have become increasingly expensive to build, and new construction techniques are constantly sought. One important new method, which came into use in the 1980s, is called roller compacted concrete, or RCC, which applies embankment dam construction techniques to concrete gravity dams. The idea of RCC is to build the dam as a single, jointless monolith. The concrete is mixed on site and then delivered to the dam by trucks, spread by bulldozers and then compacted with heavy, vibrating rollers.

Although the wall of the dam is perhaps the highest-profile aspect of construction, many additional elements must also be built, including spillways and water outlets. If the dam is designed to generate hydroelectric power massive sloping water tubes, called penstocks, are embedded in the wall during the placement of the concrete monoliths. At the base of the dam, or deep within the abutments, space to house facilities for power-generating turbines is carved out.

RESERVOIRS OF POTENTIAL

Meanwhile, behind the dam, a related task has been under way since the earliest diversion days of the project. The reservoir basin, once perhaps a forest or a desert or even a string of settlements, must be cleared—whole towns dismantled and moved, trees chopped, wildlife resettled into new habitats—to prepare for its new life as an artificial lake. As construction proceeds, the dam grows higher and can begin to hold back water. Behind it, the dammed-up water begins to fan out slowly across the terrain—well before completion of the massive structure. These artificial lakes typically are huge, sometimes taking more than five years to fill. The reservoir of the Kariba Dam between Zimbabwe and Zambia, for example, is more than 1,000 square miles in area.

Water in the reservoirs may be used for other purposes including the irrigation of vast agricultural areas, and sometimes for recreation. And although storage dams work on one level by simply holding back water in the reservoir, the structure has other roles, such as flood control and the generation of power. Hydroelectric dams work on a level rarely seen by the millions of people who benefit from them.

At a hydroelectric dam site, water from the reservoir is channeled through the embedded huge steel penstocks, that are sometimes 50 feet in diameter and weigh hundreds of tons each. Water plunges through these sloping tubes—with a vertical drop of anywhere from hundreds to more than a thousand feet—toward a powerhouse below. Gates and valves are installed at strategic points within the water-flow passages to control the intake of water and to isolate equipment for repair and replacement. Powerhouses deserve their name. These huge buildings, usually located at the base of the dam, house some of the largest pieces of equipment produced by man—huge turbines and generators. Building powerhouses inside the abutments requires astonishing feats of construction. One of the powerhouses of the La Grande 2 Dam at James Bay is 450 feet below the surrounding terrain, necessitating a mammoth underground excavation. The machine hall, where the electricity-producing generators are stored, is a third-of-a-mile-long cavern fed by six 650-foot-long penstocks.

When the falling water reaches the powerhouse, it is channeled through openings directly to the turbines—huge propeller-like devices that can weigh as much as 200 tons. The concept is simple: The turbine, powered by the water pressure, spins at 200 to 300 rotations a minute, thereby turning a shaft linked to an electricity-generating rotor. The rotor consists of a series of electromagnets encased in coils of conductive material. As the rotor turns, it induces electric currents that are collected in conducting ducts called busbars. The output from a single generator can be from tens to hundreds of thousands of kilowatts per hour. Each generator comes equipped with a governor that holds the turbine rotation speed constant by controlling the water flow. From the generator the electricity travels via the busbars to transformers which are designed to boost the voltage. Each generator is assigned multiple transformers for this task; the voltage may be stepped up between 20 and 50 times in the space of a second.

GENERATING POWER

To generate electricity, water is channeled from a storage dam's reservoir into huge pipes, called penstocks, that run through the dam. In the example of a powerhouse shown at right, based on the Hartwell Dam on the Savannah River near Hartwell, Georgia, the water (blue arrows) flows through the 24-foot-diameter penstocks at a rate of two to three million gallons per minute. When the water reaches the powerhouse, its flow is directed through openings on the inside wall of a spiral scroll case and onto the blades of the huge 91,000 horsepower, 60-ton turbine wheel, causing it to turn. The water then exits by way of a draft tube into the tailrace. The shaft connected to the turbine wheel turns the generator's armature, or rotor, between strong magnets to produce a current of up to almost 14,000 volts. Transformers boost the voltage to 230,000 volts before it is sent via transmission lines to power companies for distribution to individual customers.

1. Penstock
Conducts water from the reservoir, through the dam and to the power-house. The steel penstock may have a valve near the turbine to stop water flow.

2. Scroll case
Circulates water from the penstock and directs it to the turbine wheel. Water passes through the spiral-shaped steel tube and through openings in its inner wall to the turbine wheel.

3. Turbine wheel
Turns as water leaving the scroll case is driven against its blades.

4. Shaft
Connects the turbine wheel to the generator rotor.

5. Generator
Inside the generator, the shaft turns a coil of wire, called an armature, between powerful magnets. This produces electric power.

6. Busbar
Routes electricity from the generator to the transformer.

7. Transformer
Boosts the voltage of electricity from the generator.

8. Transmission tower
Supports high-voltage transmission lines running from the transformer.

9. Draft tube
Carries water that has passed through the turbine to the tailrace.

10. Draft tube gate
Closes to keep water from tailrace from entering the draft tube during servicing or maintenance.

11. Tailrace
Releases water from powerhouse.

Most large powerhouses contain between four and twelve such generating units located in a hall of cathedral-like proportions. Generator halls are serviced by gantry cranes that raise and lower rotors and turbines and carry out maintenance. The brain of a powerhouse is the control room. Where once the machines featured switches, meters and gauges, now video display screens and computer systems are the norm. Operators in the small, often windowless chambers, check the reservoir level, read water volume in the penstocks, check the turbine speed and the volts each generator is producing. Like all dams, a hydroelectric facility requires monitoring, regular servicing and strict regulation of water.

If there is a secret to the success of dams, it lies in their complex relationship with the natural world. Dams work because physical laws are consistent and trustworthy. Though a large dam must initially harness a river and cover the land behind it with pent-up water, soon after completion the structure participates in natural rhythms, storing resources, irrigating soil and generating energy.

TUNNELS

Surveyors, arriving at a tunnel construction site to chart the day's progress, are confronted with pops, creaks and groans and eerie reverberations as the earth settles against the tunnel's freshly placed supports and lining. Even at night, long after the workers have gone home, a tunnel under construction can be a noisy environment. The sounds are created as hard rock or soft ground attempts to move into the void created by the advancing tunnel. The noises are a measure, too, of the stresses tunnels constantly endure. Unlike other large structures that exert pressure on a supporting foundation, a tunnel shoulders massive loads itself—in essence it is one large foundation. Tunnels, therefore, must be constructed so that the loads are effectively and unfailingly withstood—whether the stress comes from the sheer compression of the surrounding ground, the hydrostatic pressure of water from all directions, or the live load brought to bear by the trains, automobiles or water that will run through them.

Of all large engineered structures, tunnels are probably the least obvious and the most taken for granted. Skyscrapers, bridges, dams and stadiums are bold statements that become part of city skylines or create new ones. Tunnels, on the other hand, are out of sight and are not even expected to look good; words such as "soaring" or "monumental" are rarely applied to these subterranean or subaqueous structures. Only occasionally is a tunnel thrust into the limelight, as in the case of the tunnel built under the English Channel—known as the Chunnel—or when a new metropolitan subway line opens up. But tunnels have been proving their usefulness since at least 2160 B.C., when the Babylonians constructed one to enable their kings to cross the river Euphrates in secrecy. Its builders first had to divert the river and then build a tube of bricks 3,000 feet long, which they waterproofed with asphalt; when its waters were released, the river completely covered the ingenious structure.

A passenger train approaches the 709-foot-long Landwasser Tunnel in Graubünden, Switzerland. Completed in 1903, the tunnel, and the viaduct leading up to it, form a section of the famous "Glacier-Express" train route that runs between Zurich and the mountain resort of St. Moritz.

Generally speaking, there was never a great need for tunnels until transportation networks in Europe and North America began to expand in the late 18th and 19th Centuries. The first transportation tunnels linked canals and other waterways. Then, as railway tracks began to criss-cross new and far-flung territories, trains were obliged to pass through increasingly difficult and hilly terrain. Tunnels were needed to flatten out routes because locomotives had little hill-climbing ability. And as the populations of large cities grew, a greater need for more efficient and less intrusive underground transportation and utilities networks grew accordingly.

Today, tunnels link islands to each other and to their respective mainlands. Tunnels bring fresh water to cities, often relying on the force of gravity to sustain the flow from distant reservoirs and deliver utilities such as natural gas, electricity and cable television to urban homes. Sewer tunnels drain away billions of gallons of society's waste every day. Subway tunnels whisk passengers along hundreds of miles of track in a score of cities around the world, while trains and cars routinely pass through tunnels under riverbeds. Many engineering projects under construction, including bridges and dams, rely on temporary tunnels to divert water away from construction sites.

TUNNEL BASICS

This illustration of a typical automobile tunnel shows the basic features of many tunnel types. One or more linings, which are installed against the earth or rock walls, are made of concrete or steel and serve to support the structure. A finished tunnel requires a drainage system, utility and equipment-storage niches, interior lighting, a power supply and systems for fire protection, ventilation and communication.

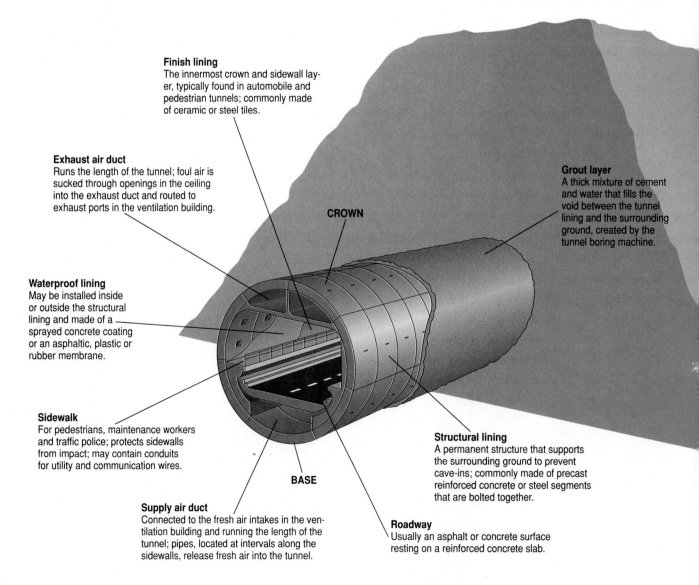

Finish lining
The innermost crown and sidewall layer, typically found in automobile and pedestrian tunnels; commonly made of ceramic or steel tiles.

Exhaust air duct
Runs the length of the tunnel; foul air is sucked through openings in the ceiling into the exhaust duct and routed to exhaust ports in the ventilation building.

Waterproof lining
May be installed inside or outside the structural lining and made of a sprayed concrete coating or an asphaltic, plastic or rubber membrane.

Sidewalk
For pedestrians, maintenance workers and traffic police; protects sidewalls from impact; may contain conduits for utility and communication wires.

CROWN

Grout layer
A thick mixture of cement and water that fills the void between the tunnel lining and the surrounding ground, created by the tunnel boring machine.

Structural lining
A permanent structure that supports the surrounding ground to prevent cave-ins; commonly made of precast reinforced concrete or steel segments that are bolted together.

Supply air duct
Connected to the fresh air intakes in the ventilation building and running the length of the tunnel; pipes, located at intervals along the sidewalls, release fresh air into the tunnel.

BASE

Roadway
Usually an asphalt or concrete surface resting on a reinforced concrete slab.

TUNNEL TYPES

Most large engineered structures are defined by the way they work, how they bear loads, and what they are used for. This is also true for tunnels, but just as important a factor in defining tunnel type is the nature of the ground they must go through. There are two broad categories: soft and hard ground tunnels. For a tunneler, soft ground is earth that does not support itself after it is excavated, and that therefore must be shored up almost immediately. Soft ground is typically composed of soil, but also includes rock so fractured that it behaves like a soil. Soft ground's most important characteristic is its "stand-up time"—the period during which freshly excavated ground will support itself before it needs to be reinforced. In the case of loose sands and silts, this may be only a matter of minutes or seconds. If water is mixed in with the soil, there may be no stand-up time at all. Silts and clays tend to be easier to excavate because their stand-up times may be hours long. Traditionally, soft ground tunnels were dug with picks and shovels; today, they are usually advanced with elaborate tunnel boring machines, known in the trade as TBMs, or moles, which dig, remove the loosened ground, and install reinforcement in an ongoing process.

Though hard ground is more self-supporting than soft ground when excavated, stand-up time is still a factor. Huge masses of solid, intact rock have high strength and may be permanently self-supporting when a tunnel is driven through them. More often, however, hard rock is not a contiguous mass, but is made up of blocks separated by large fractures called faults, and smaller cracks called joints. These discontinuities mean that the weight of individual large rock blocks may put extreme pressure on a tunnel. As well, the cracks and faults may be filled with other weaker materials, and sometimes water. As with soft ground tunnels, hard ground tunnels are normally given additional support after excavation.

Tunnels that are built through hard rock traditionally are advanced by drilling forward, deep into the rock, and setting explosions. Afterward the debris is cleared away manually or by machine. Some of the world's most famous mountain tunnels, including the four-and-three-quarter-mile-long Hoosac Mountain Tunnel in Massachusetts and the eight-and-a-half-mile-long Mont Cenis Tunnel through the Alps between France and Italy, travel through hard rock.

Two other tunnel types, normally advanced through soft ground, also are defined by the method of digging or construction used to build them. A cut-and-cover tunnel is first excavated as a trench from the surface. The floor, walls and ceiling of the tunnel are built in the exposed trench, which is then covered with backfill. A prefabricated subaqueous tunnel, called a sunken tube, is built by sinking ready-made steel or concrete sections into an excavated trench at the bottom of a body of water and then connecting the sections. Increasingly, however, improved construction technology has blurred some of these general distinctions of soft and hard ground tunnels. New TBMs, for example, now grind through some types of hard rock as if they were soft ground.

In cross section, tunnels come in a variety of shapes. Cut-and-cover tunnels, which are located near the surface and are normally rectangular, present a horizontal roof to the relatively light weight of the ground pushing down from above. Sunken tubes may be circular or rectangular in cross section depending on what they will be used for. The many tunnels which are advanced deep beneath the

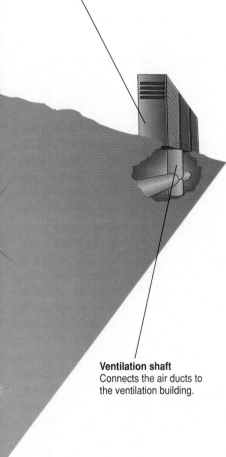

Ventilation building
An above-ground building that contains the fresh air intakes, exhaust ports and the fans of the ventilation system; also houses the control room, workshops and equipment, and employee facilities.

Ventilation shaft
Connects the air ducts to the ventilation building.

surface of the earth—whether blasted or bored—are always circular, semicircular, or at least arched at the top. A curved roof exploits the natural strength of the arch and reliably withstands the heavy downward thrust of the great amount of rock or earth above. A tunnel with curved walls also can endure the horizontal compressive push that comes from the ground pushing in from the side; TBMs, which are in increasing use, bore forward in a path with a circular cross section. Tunnels that will be used regularly by people or machinery have a flat bottom upon which tracks or roadways are built. The tunnel may be elongated in cross section to provide more horizontal room to run multiple tracks or roadways. On the other hand, tunnels that serve as water conduits normally are fully circular in cross section; the shape enhances efficient water flow. Hard rock tunnels traditionally have had arched roofs, with vertical or horseshoe-shaped sidewalls.

UNDERGROUND WINDOWS

Given the importance of ground type, it is not surprising that before construction of a tunnel begins, its builders need a complete picture of the geological world through which the tunnel will travel. Extensive site investigation determines the tunnel's route, defines what type of equipment is best suited for the job and predicts some of the hazards the tunnelers will meet as the structure advances. But even though tunnelers take every precaution to reduce risk, they still expect the unexpected. No amount of preliminary investigation can completely eliminate unforeseen problems, and the history of tunneling is darkened with nightmarish stories of floods, cave-ins and explosions.

The first stage of the site investigation may be remote sensing, which usually entails an aerial photo or satellite image interpretation of the proposed route. These pictures may hint at important geological formations beneath the surface. A trained interpreter even can draw conclusions about the nature of the ground beneath from the types of vegetation present on the site. Infrared sensing and photography also may be employed in the investigations. Infrared technology detects changes in thermal radiation, or heat, and helps to identify zones with groundwater. Where the tunnel will be underwater, different preliminary techniques may be used. During the site investigation phase for Japan's Seikan Tunnel—a 34-mile-long route under the Tsugara Strait between the islands of Hokkaido and Honshu—underwater listening devices called geophones were lowered beneath the surface. When dynamite was exploded near the seabed, the geophones monitored the sound waves that bounced back, revealing hidden geologic boundaries between rock layers, or strata.

But the most revealing—and reliable—exploration technique is still to drill bore holes through the soil and rock. The process involves powerful mechanical drills that can bore many hundreds of feet into the ground, yielding a long, circular core sample, or profile, of the soil strata or rock beneath. Normally, the boreholes extend to a depth that exceeds the intended depth of the tunnel—by a margin of about twice the stucture's diameter. For a long tunnel, hundreds of borings may be made along the proposed route, each, in effect, a single brush stroke in what eventually will be a composite picture of the subterranean world. The geologists will know if the ground is hard or soft, or both, where the soil or rock is consistent and where there are faults or even caverns. Rock cores reflect discontinuities in the rock and

Illustrated at right in cross section are the four basic tunnel types: arched, rectangular, circular and sunken tube. The shape of each tunnel is determined by a number of factors: its purpose, which may range from transportation to water conveyance, and the construction technique used for excavation, which depends on the type of ground. All tunnel types are engineered to withstand the compressive stress created by the surrounding ground, ranging from hard rock to soft soil, the hydrostatic force of subterranean water and the live loads of the automobiles, trains or water that travel through them.

Arched tunnel
This type of tunnel is commonly used to route car and train traffic through mountains. Its shape is determined by the terrain—the arch is efficient at withstanding the compressive forces of the rock mass above. Excavated by drilling and blasting with dynamite, the tunnel may require a structural lining to support the surrounding rock.

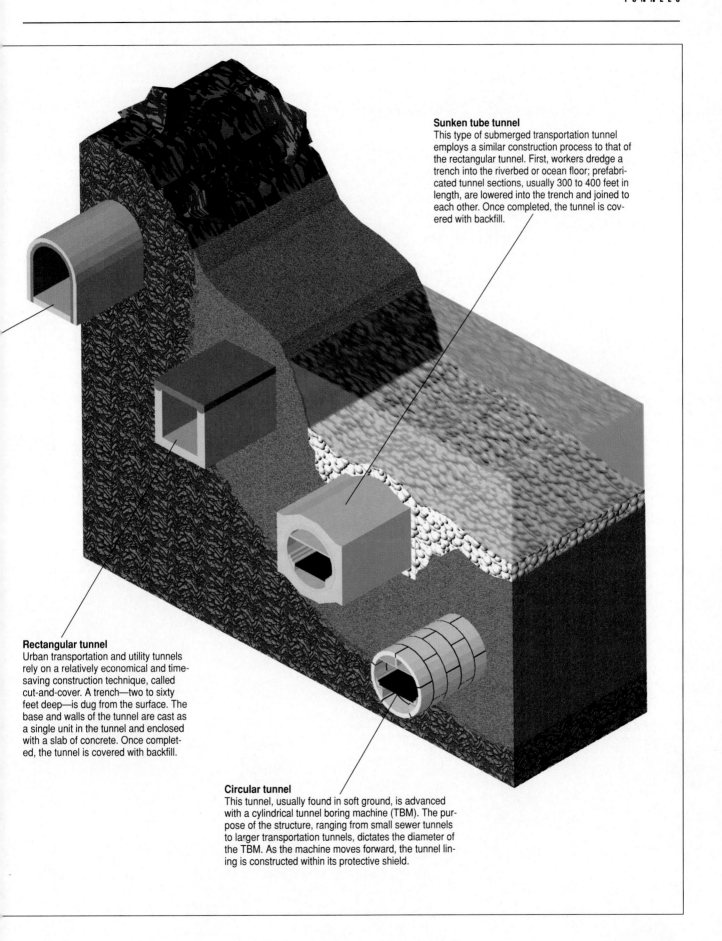

Sunken tube tunnel
This type of submerged transportation tunnel employs a similar construction process to that of the rectangular tunnel. First, workers dredge a trench into the riverbed or ocean floor; prefabricated tunnel sections, usually 300 to 400 feet in length, are lowered into the trench and joined to each other. Once completed, the tunnel is covered with backfill.

Rectangular tunnel
Urban transportation and utility tunnels rely on a relatively economical and time-saving construction technique, called cut-and-cover. A trench—two to sixty feet deep—is dug from the surface. The base and walls of the tunnel are cast as a single unit in the tunnel and enclosed with a slab of concrete. Once completed, the tunnel is covered with backfill.

Circular tunnel
This tunnel, usually found in soft ground, is advanced with a cylindrical tunnel boring machine (TBM). The purpose of the structure, ranging from small sewer tunnels to larger transportation tunnels, dictates the diameter of the TBM. As the machine moves forward, the tunnel lining is constructed within its protective shield.

Beneath City Streets

The thicket of landmark skyscrapers that makes up Manhattan—the Empire State Building, the Chrysler Building and the World Trade Center among them—reaches for the sky over an area of only 22 square miles. Hidden beneath the buildings, and a thin outer skin of asphalt, is an equally impressive landscape that plummets to a depth of 76 stories.

The island is made of hard rock—called Manhattan schist—that supports the maze of tunnels carrying electricity, gas, water and heat. Layer under layer, the vital arteries of the infrastructure descend to a depth of 800 feet. The uppermost slice—only two feet deep—contains more than 11 million miles of telephone, electric, streetlight, fire alarm and cable TV wires. The world beneath the pavement bottoms out with water-supply tunnels. Sandwiched between is the subway system, ranging in depth from 3 to 180 feet, with its 63.8 miles of underground track, and older water tunnels.

Excavation for these cable, main, sewer and subway tunnels yields a trove of buried structures. There are pneumatic tubes, once used to whisk letters swiftly through the city, a retired City Hall subway station, private tunnels to Prohibition speakeasies and evidence of Manhattan's first water supply pipes—hollow tree-trunks—abandoned in the early 1800s.

Other than to catch the subway, few New Yorkers venture into the bowels of their city; but for those who toil below it, the day's work consists of keeping sewage and millions of gallons of potable water flowing and constantly maintaining the infrastructure of this underground world.

A row of giant valves (left) *stretches into the distance in a tunnel 250 feet under New York City. The valves, housed in their enormous 620-foot-long chamber, control the water's flow and destination. The inverted profile of the Empire State Building* (right) *puts into perspective the depth of Manhattan's underground world.*

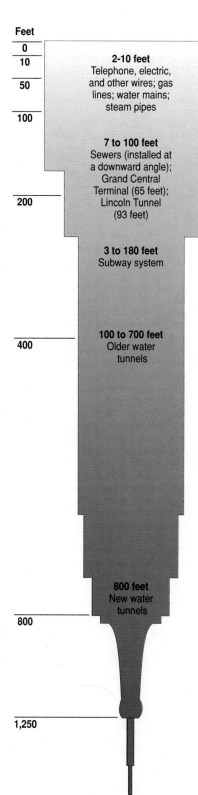

Feet

0
10
50
100

2-10 feet
Telephone, electric, and other wires; gas lines; water mains; steam pipes

7 to 100 feet
Sewers (installed at a downward angle); Grand Central Terminal (65 feet); Lincoln Tunnel (93 feet)

200

3 to 180 feet
Subway system

100 to 700 feet
Older water tunnels

400

800 feet
New water tunnels

800

1,250

give a good indication of the stability of the larger rock formations that will be encountered during tunneling; analysis of the ground gives a measure of its strength and composition.

Tests done in the drill holes can also indicate the presence, the rate of flow and pressure of groundwater. Whether in rock or in soil, water is a particular problem for tunnelers. It can seep into an excavated tunnel from any direction; its pressure in sand can produce a dangerously flowing quicksand mixture. In hard rock, water may flow through fractures and sometimes lubricates the joints in the rock, weakening its ability to stand unsupported. Groundwater has, in the past, flooded tunnels, threatening or taking lives; these days, however, with increased safety precautions, its presence in unexpected quantities usually leads to cost overruns rather than catastrophe.

Investigation of the ground does not stop once tunnel construction is underway—core samples may continue to be taken throughout the construction phase. In many tunnels, small-diameter pilot tunnels, also known as drifts, are driven ahead of the main tunnel, yielding vital advance information about the ground ahead. Instruments called extensometers measure even minute amounts of deformation caused by tension, compression, bending or shear in the rock. Lowered into boreholes or positioned in the pilot tunnel these sensitive instruments evaluate the strain in rock or the movement and settling of the ground after excavation. Other sensors measure pressure from the surrounding ground on the lining or supports of a tunnel. The entire investigation may take years. Testing for the Seikan Tunnel, for instance, began in 1946. Eighteen years later, the first shafts and exploratory pilot tunnels were advanced; the tunnel was completed only in 1988.

TUNNEL DRIVING

Once a route has been chosen, and the ground conditions plotted, the construction of a tunnel begins. Many shallow tunnels—typically those built under streets—are constructed with the cut-and-cover technique. Tunnels that lie about 35 feet below the surface are often dug from the surface rather than bored beneath it. The process, commonly used for the construction of subway systems, is perfectly straightforward. With traffic detoured and utilities supported or rerouted, a wide trench is evacuated to the required depth; a base, usually of reinforced concrete, is then laid in the trench and the tunnel is built upon it. Once the concrete has hardened, the previously excavated soil and gravel is shoveled back into the trench. The backfill protects the tunnel from any pressures exerted above ground.

The cut-and-cover technique is also the starting point for sunken tube tunnels. In this case the trench is dug or dredged in the river, lake or ocean bed, usually by a hydraulic dredge or clamshell dredge, mounted on a barge. Then prefabricated watertight steel or precast concrete tube sections, each end of which is sealed with a bulkhead, are floated to the tunnel site on barges and sunk into position. The sections, usually 300 to 400 feet long, are joined together with collar-like rings that are bolted together and which may be filled with concrete. Normally, divers perform this work. To protect the tunnel, the trench is filled with soil that is put in place by a clamshell digger or by a long tube, mounted on a barge. Finally, a layer of rock is placed along the surface of the filled trench. Inside the tunnel, the bulkheads are removed and the tunnel prepared for its use.

A tunnel drilling rig, called a jumbo, works at a tunnel face. An operator, working in the cab, uses a joystick to move the two outstretched booms into position on the surrounding rock. The drills, mounted to the end of each boom, have tungsten-carbide tips that can bite as far as 16 feet into the rock face, creating holes that will be charged with an explosive material.

HARD ROCK TUNNELING

The essentials of advancing a tunnel underground, whether through soft or hard rock, are more or less universal and have remained unchanged for centuries. The challenge still is to drive the tunnel forward while disposing of broken rock and soil—the muck—by moving it backward. Tunnelers must work swiftly to support the opening before it has a chance to collapse under the stress of the ground around the new excavation.

Advancing a hard ground tunnel is typically done by the drill and blast method. Here, a series of holes are drilled in the rock face and charged with explosives, which are then ignited. After the subsequent blast, the fumes are exhausted through a ventilation pipe, the broken rock removed and temporary supports placed against the exposed rock. Drilling takes up a major portion of the cycle, known as a round. At first, drilling was done manually with sledgehammers and long drills by specialists known as steel-driving men; the most famous is the legendary John Henry, who drilled 14 feet through hard rock in 35 minutes while laboring in the Big Bend Tunnel for the Chesapeake and Ohio Railroad in 1870. But he was one of the last of a rugged breed. At about that time, the drilling process was speeded up immensely with the introduction of pneumatic drills; pneumatic and hydraulic drills are still commonly used. In both cases the drills are equipped with tungsten-

carbide tips that penetrate the rock with both rotary and percussive—or hammering—action. Traditionally, a single worker would operate a drill mounted on one of a number of decks that formed a multilevel platform called a jumbo. The advantage of a multideck jumbo was that it would allow many workers to drill at different levels of the rock face simultaneously. The entire apparatus would have to be moved laboriously back before the blast, then dragged forward again for the next cycle. Today, the old jumbos have been replaced by motorized vehicles, which are still called jumbos. Here the drilling is done entirely by drills mounted on long mechanical arms, known as booms, operated by one or two workers in the cab of the tractor-like vehicle. The holes are usually drilled no further than about 3/4 the diameter of the tunnel, the maximum safe distance for a single advance. When the drilling is complete and the holes charged with explosives, the jumbo is backed up out of harm's way until after the explosives are detonated, and then pushed forward again for the next cycle of drilling and blasting. While the drilling is in progress, water, ventilation and power lines are extended forward to the working area of the advancing tunnel.

The explosion must go off in such a way that the blast comes back toward the face of the tunnel; otherwise it will be absorbed by the surrounding rock and the face will not crumble. The rubble should be thrown back into the tunnel with as little damage as possible to the surrounding rock. The blasting pattern typically consists of a series of holes: trim holes, drilled almost parallel to the tunnel and located around its perimeter; intermediate holes, called relief holes, drilled into the face at a slight angle; and an innermost group, called cut holes, drilled at sharp angles. Explosives are positioned in each of the holes; explosives used in the cut holes are lighter to avoid damaging the rock around the tunnel. The timing of the explosions at each hole type is staggered—the whole process may take 15 seconds. The cut hole explosives go off first and create a space into which the rubble from the relief-hole and trim-hole blasts will go.

Accuracy in selecting the depth, angle, alignment and pattern of drilling still depends mostly on the skill and intuition of human operators. In Europe, drilling and blasting is often more mechanized; in some cases, computers in the jumbos calculate the appropriate drilling angles and depths. Now, engineers are at work on machines that will drill and charge holes in a single operation controlled by the tunnelers from the cab of the jumbo.

UNDERSTANDING GEOSTRESS

As they advance underground, tunnelers must always take into account the *in situ* geostress—that is, the amount and type of stress in the rock where the next round of blasting will take place. Investigations by engineers skilled in the field of rock mechanics help determine what kind of response the rock will have to different types of excavation or the intensities of blasts. The goal is to blast out a predictable amount of rock in a predictable pattern. But because rock is not necessarily a solid mass, individual blocks of varying size may push in on the walls of a tunnel. The resultant geostress may occur at different spots along the tunnel walls, and with varying intensity.

The fastest way to drill and blast a tunnel in rock is to use the full-face method, in which the full diameter of the tunnel is driven forward at each round of blasting.

However, if the rock is unstable or unpredictable and the tunnelers believe that there may not be enough stand-up time to support the tunnel, other techniques are used. One method, in which limited portions of the rock are removed one at a time, is called top-heading. Here, the upper, arching part of the tunnel is blasted first, leaving in place a bottom portion, known as the bench, which extends back toward the workers. The bench is then blasted away separately. In the drift method, two or more small tunnels, called drifts, are driven at the side of the projected tunnel, parallel to and ahead of the main bore. Steel or timber posts, or concrete abutments for the arch supports, are installed in the drifts and only then is the balance of the tunnel face opened up.

After the explosion, smoke is exhausted through a pipe that has been forwarded along with the tunnel. The rock removal, or mucking process, is a low-tech affair of conveyor belts, power shovels, rail cars and sometimes trucks. In small tunnels, workers with shovels still do the job.

Keeping the tunnel on the proper line as it advances is a fairly simple job of surveying, but it must be done constantly. Traditionally, the alignment was assured by a surveyor who used a system of levels and plumb bobs suspended from the tunnel roof, but they were susceptible to movement due to settlement. Now tunnelers commonly use a laser beam system in which an intense beam of light replaces the plumb bobs. The laser gun is mounted in a protected position in the

Lining

Shell

Body

Foot

The Inspirational Shipworm

In 1815 Marc Brunel picked up a piece of worm-eaten timber from a ship under repair. The famed engineer realized that the creature who could do so much damage to a ship might provide the answer to a vexing technical problem: how to drive the first pedestrian tunnel beneath the River Thames without setting off dangerous cave-ins. The shipworm extends a suction organ, called a foot, through the two valves of its shell. With the foot anchored to the burrow, the mollusc twists its body up to 180°, rotating the rasp-like cutting edges of its shell, producing tiny shavings of powder. It disposes of the powder by eating it and then excreting a substance that forms a hard lining in the tiny tunnel. With the shipworm as his inspiration, Brunel designed the first shield which used a hard hood to provide protection for advancing tunnelers. Like the worm, the excavated material was moved back through the shield and disposed of. The original shield was a precursor of more modern shields and today's massive tunnel boring machines.

Boring through a piece of wood at left are a number of *Teredo navalis*, the species of mollusc commonly known as the shipworm.

MAN-MADE MOLES

Directed on course by means of laser beams, tunnel boring machines (TBMs) called moles are custom-built for chewing through ground ranging from wet muck to solid rock. Illustrated below in cross section is a 21-foot-long soft-ground TBM that weighs 250 tons. With motors supplying 1,020 horsepower, this TBM bores a diameter of 18.75 feet, almost enough clearance for a two-story house.

The machine is propelled forward by the pushing force of 18 hydraulic jacks. As it moves, the TBM's cuttinghead rotates and its teeth rip into the ground. Stabilizing fins can be projected outward from the TBM into soft ground to prevent any rolling motion if hard ground is encountered. The excavated material, called muck, is fed from the cuttinghead onto a conveyor inside the TBM that carries the unwanted soil out the back to a waiting muck car in the tunnel.

During each tunnel-boring cycle, the TBM's jacks take approximately 15 minutes to propel the apparatus a distance of 6 feet. Then burrowing is suspended for an hour or so while the freshly excavated section is lined. With the cuttinghead stopped and the hydraulic jacks retracted, a set of lining segments of precast, reinforced concrete is installed within the tail shield of the TBM using a segment erector arm. Then a pushing ring distributes the pressure of the hydraulic jacks on the new lining during the next cycle.

Corrections in the course of the TBM can be made by adjusting the hydraulic jacks, altering the pressure they put on this ring that in turn pushes against the most recently placed lining segments. Meanwhile, a train shuttles the muck out of the tunnel, returning to the TBM with grout, more lining segments and empty muck cars. As the next cycle begins and the TBM is again propelled forward, a sealing grout is pumped into the gap left around the lining segments and more track is laid to further extend the reach of the train.

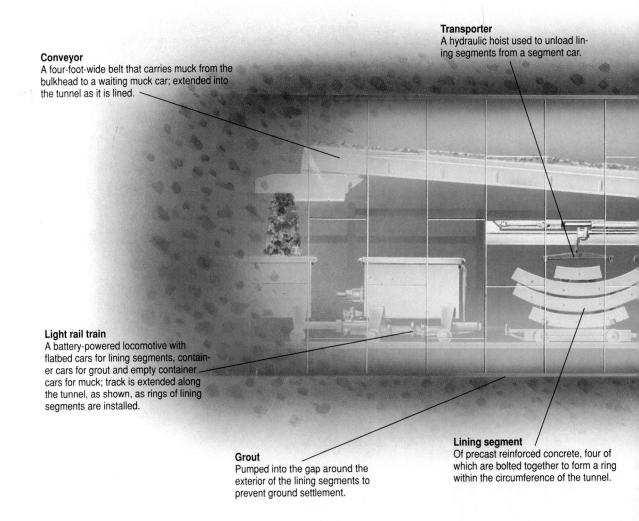

Conveyor
A four-foot-wide belt that carries muck from the bulkhead to a waiting muck car; extended into the tunnel as it is lined.

Transporter
A hydraulic hoist used to unload lining segments from a segment car.

Light rail train
A battery-powered locomotive with flatbed cars for lining segments, container cars for grout and empty container cars for muck; track is extended along the tunnel, as shown, as rings of lining segments are installed.

Grout
Pumped into the gap around the exterior of the lining segments to prevent ground settlement.

Lining segment
Of precast reinforced concrete, four of which are bolted together to form a ring within the circumference of the tunnel.

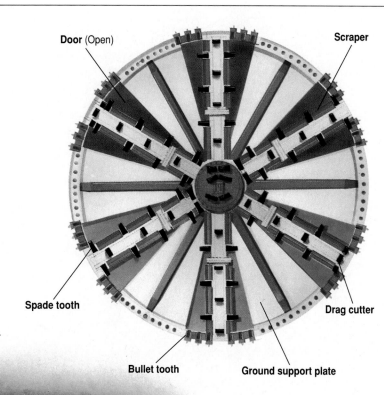

Door (Open) **Scraper**

Spade tooth

Bullet tooth **Ground support plate** **Drag cutter**

CUTTINGHEAD

Constructed of high-strength steel, the cutting-head of the TBM is shown face-on at right. With its ground support plate pressed against the face of the tunnel to keep it from collapsing, the cuttinghead rotates and bites into the ground. Its spade teeth dig into soft soil; its drag cutters, with their tungsten-carbide inserts, rip into soft rock, carving grooves and shearing material between them. The bullet teeth of the cuttinghead cut a clearance, or kerf, wide enough for the TBM to pass. Excavated ground, called muck, is loaded into the cuttinghead by the scrapers and directed through operator-controlled doors into the TBM.

Propulsion cylinder
One of 18 containing hydraulic jacks, each capable of extending almost six feet and pushing up to 125 tons; they provide forward motion by pressing against the ring installed on the most recently placed lining segments.

TUNNEL **TAIL SHIELD** **STATIONARY SHELL** **FORWARD SHELL**

Cuttinghead
Rotates on the bulkhead, burrowing into the ground and directing excavated material, called muck, through its doors.

Control console
A centrally positioned panel of hydraulic controls, operating and monitoring gauges. An automatic shutdown is triggered when sensors detect a hazard—dangerous gas in the ground or a rolling motion of more than 10° from horizontal, for example.

Segment erector arm
hydraulic arm used to construct each ring of lining segments.

Pushing ring
A steel ring of the same diameter as a set of lining segments that distributes the pressure of the propulsion cylinders on them.

Stabilizer fins
Projected outward into soft ground to prevent rolling motion when the cuttinghead encounters hard ground.

Bulkhead
Interior face behind the cuttinghead; open at the top for muck to enter and land on the conveyor.

tunnel, well back of the excavating activities. It is aimed at a grid on the rock face, or at targets on the excavating equipment, and determines the line and level of the advancing tunnel. Once the laser has established these two factors, its beam is immune from any disturbance; the tunnelers have a clear idea of alignment at all times by looking at the beam.

SUPPORTING THE ROOF

Except in the most solid rock formations, the roof of the tunnel must be braced as soon as possible after excavation. To do this, tunnelers need to establish an initial, temporary support. Formerly, massive columns and beams, roughly following the shape of the tunnel, were installed. The gaps between the timbers and the curved rock walls were filled with interspaced wooden blocks or packed with crushed rock. More recently, curved steel columns, or ribs, that match the contour of the tunnel are used instead of wood for primary support. The advantage of ribbing is that it can be adjusted periodically to the levels of rock stress as they build. Depending on the level of the geostress, the ribs may be installed every five feet; the spaces between are covered or partially covered with horizontal steel or wooden supports, called lagging.

Today, tunnelers use innovative and more cost-effective techniques to provide just enough reinforcement for the rock to become self-supporting. One technique is to use long steel bars called rock bolts that are driven into holes drilled at right angles to the tunnel roof. The bolts, which range from .75 to 1.5 inches in diameter, act like reinforcement bars in concrete, adding both shear and tensile strength to fractured rock. The bolts are typically ten feet long; some are hollow, and a grout mixture of cement and water is injected through them to further stabilize the rock. Rock bolts may also be used in conjunction with shotcrete—a concrete mix that is sprayed onto the crown and sidewalls of the excavation to provide immediate support. The standard concrete aggregate and cement for shotcrete may be mixed with steel or fiberglass reinforcing fibers and can be used with or without conventional steel reinforcing ribs. A construction technique called the New Austrian Tunneling Method (NATM) has been popular in Europe since its introduction in the 1950s, and is now gaining wide acceptance in North America. NATM is characterized by the immediate application of shotcrete after blasting for early support of unstable rock or mixed faces of earth and rock. The shotcrete may be applied to the vulnerable arch of the tunnel's ceiling even before the debris covering the lower part of the tunnel has been removed. The NATM depends on immediate and frequent testing to measure whether or not the stress exerted by the rock is increasing, thereby possibly deforming the concrete. The results tell the tunnelers whether more shotcrete or bolts are needed and allows a much more efficient use of materials. The goal is to install only as much support as is necessary.

For rock tunnels that need additional support, a permanent inner lining of concrete is pumped or poured into wooden or steel sectional forms that are built after the primary support has been established. Precast lining sections of concrete or steel can also be made outside the tunnel and transported to the face. In locations where the geostress is spread evenly along the tunnel walls, the lining is usually not reinforced. Where the stress is not consistent on top of the tunnel, the amount of reinforcement is increased.

SOFT GROUND TUNNELING

The stresses exerted by hard rock blocks are of little concern to soft ground tunnelers. But tunneling in soft ground, especially in the presence of water, presents its own special difficulties. The most pressing is the certainty of collapse. To counteract this, soft ground tunnels are usually bored with some form of shield to protect workers from cave-ins. Invented in the late 1820s by the Anglo-French engineer Marc Brunel, the earliest shields were little more than movable platforms that put a temporary roof over the heads of workers as they labored at the tunnel's face. Depending on the stand-up time, the workers would pick at the face through small openings in a rectangular shield in front of them. As the tunnel advanced, the shield was inched forward by screw jacks braced against the tunnel walls. The muck was taken out through the apparatus while workers immediately behind the shield installed a brick lining around the entire diameter of the tunnel. A timber and iron

platform shield allowed a tunnel to be driven under the River Thames in London, the first successful tunnel to be bored under a body of water. Completed in 1843, the brick-lined, 1,200-foot-long Thames Tunnel is still used by the trains of the London Underground. But shields truly came into their own in 1874, when another tunnel under the Thames was driven using a circular version. Developed by the tunnel engineer James Henry Greathead, the shield was driven forward into the soft ground like a cookie-cutter. Workers were protected under the hood of the shield; the face of the shield was divided into segments which could be closed quickly in case of an inflow of water or muck. When seeping water was a problem, the digging would take place in a sealed chamber filled with compressed air. The air pressure kept the water at bay, in a system very similar to the compressed air—or pneumatic—caissons used to dig the underwater foundations of large bridges. A circular lining of bolted, cast-iron segments replaced the brick lining of Brunel's tunnel.

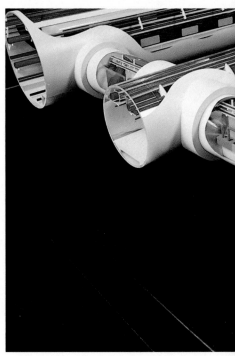

In many respects, modern tunnel boring machines—often called moles—are the direct descendants of Brunel's and Greathead's shields. These sophisticated machines simultaneously excavate and remove muck with huge rotating cutting heads, while offering complete protection from ground collapse. Within the mole workers operate hydraulic jacks which act as huge arms that position and force into place tunnel liner sections behind the cutting wheel. The segmented lining is actually constructed within the shield, so that ground pressure will not be brought to bear on the lining until the shield has been jacked forward.

There is no single standard design for a TBM; many are custom-built for one particular job and require refitting before being reused on another tunneling project. Soft ground TBMs may have an open face for use with self-supporting ground conditions, or be full-breasted—virtually closed at the cutting end to provide support against the possibility of soft ground collapse. Water or water mixed with soil is always a potential problem which is combated by various means. In some cases much of the water is pumped, from the surface, away from the tunnel route; in other instances advance borings are made to drain off the water. When water cannot be removed in advance, specialized tunneling techniques are used to ensure the safety of the tunnelers. A hydroshield is a specialized TBM with a rotating cutter head that forces bentonite slurry—a clay and water mixture with the consistency of a milk shake—under pressure onto the face of a tunnel. There the slurry performs a dual function. As the fluid pressure keeps the tunnel face from caving in, the excavated soil mixes with the slurry. The thick muck is pumped to the rear of the tunnel boring machine. When the mixture reaches the rear of the tunnel, the soil is filtered out and the slurry is pumped back through pipes to the face to be reused again and again. Sometimes the water is held back with compressed air or by being frozen, but both methods are expensive and infrequently used. A hybrid solution is the mixshield TBM that can adapt to changing soil consistency; it converts from a slurry shield to a non-pressurized, open-faced cutting machine, depending on the ground it meets.

MOLES, LARGE AND SMALL

The smallest TBMs are remote-controlled units, too diminutive to accommodate human tunnelers. They are used to bore microtunnels—an increasingly accepted and economical alternative to cut-and-cover tunnels for laying utility pipes under cities. The largest TBMs are vast machines such as the 11 monsters built to excavate the Chunnel. The longest of these moles, including its conveyor belts, track-laying equipment and other trailing gear, was more than 800 feet long, weighed 1,200 tons and cost $20 million. The jacks that forced the 28-foot-high, tungsten-carbide tipped cutting head into the chalk marl exerted a force of more than 10,000 tons. The marl, also known as Blue Chalk, proved to be a nearly ideal tunneling medium: Waterproof, 65 to 100 feet thick and 90 to 500 feet below the sea bed, it was soft enough to be scored with a fingernail, but strong enough to be self-supporting until the TBMs installed the lining sections.

The Chunnel sections were kept in alignment by a computer and laser system that constantly checked the tunnel's progress against a dozen reference points established with the aid of four Navstar satellites. Whenever the beam moved off-center, a microcomputer adjusted the TBM's on-board hydraulic jacks to steer a true course. The cutter heads rotated at up to three revolutions per minute, and under optimal conditions the TBMs progressed at a rate of 14 feet per hour. The British TBMs could remove 2,000 tons of chalk every hour—initially on internal spoil conveyors, then in railroad cars on track that was laid continuously at the rear of the TBM. At the same time, equipment and concrete liner sections were

At left, French workers on the tunnel beneath the English Channel—popularly known as the "Chunnel,"—operate a powerful hydraulic jack to press a prefabricated segment of lining into place at the top of the advancing tunnel. The Chunnel, which will be one of the longest tunnels in the world, will stretch 23.5 miles beneath the waters that separate France and England. The model below shows a section of the Chunnel. Trains, carrying automobiles and passengers, will run in opposing directions through the two outer tunnels. In the middle is a service and safety tunnel, which is connected to the main tunnels at regular intervals by passageways.

brought into the tunnel along the tracks at a rate of 500 tons per hour. The TBMs paused every five feet for mechanical arms to thrust the eight- to nine-ton liner sections into place and for workers to inject a grout mixture between the tunnel wall and the lining. The grout layer filled the voids between the lining and the wall, evening out the pressure and helping to keep out water.

Among those who work on them, tunnels are half-jokingly defined as "structures that leak." All tunnels—whether in soft ground or hard, despite grout and lining layers—will face the problem of water seepage. In some cases a thick plastic water-proof lining may be installed between the linings, or sometimes immediately against the rock walls. Most tunnels are equipped with drainage systems that collect water from inside the lining and pump it away.

THE WORKING TUNNEL

Once the basic structure is completed, the system is supplied with all the services and infrastructure it needs—train tracks, platforms and escalators for subway tunnels as well as all the more mundane services such as plumbing, drains, lighting, fire protection and communication systems.

Of all the infrastructure services, ventilation is considered the most crucial element in a working tunnel system. Even in subway tunnels used exclusively by electric trains, the heat generated by traffic, machinery and human bodies would cause temperatures to rise to intolerable levels without an adequate ventilation system. In the event of accidents, traffic jams or fires in automobile tunnels and subways, ventilation may well make the difference between survival and suffocation. Automobile tunnels also have to cope with the constant emission of carbon monoxide and other toxic gases.

The first automobile tunnel built on a grand scale in the United States was the Holland Tunnel, opened in New York in 1927; it was ventilated by 84 enormous fans mounted in buildings at either end of the tunnel. The basic role of ventilation fans is to import fresh air, keep it cool, and to force out foul air. The longitudinal ventilation system uses axial jet fans mounted inside the tunnel to amplify the natural draft produced by the traffic. This system is suitable for automobile tunnels where traffic moves mainly in one direction, although the direction of the airflow is often designed to be reversible to respond to morning and evening rush hours when the traffic flow may be reversed. In tunnels such as subways, where there is not usually enough headroom to install the fans on the ceiling, longitudinal air-flow can be generated by mounting fans in intermediate vertical shafts with access to the surface.

In a transverse ventilation system, air is supplied by external fans through openings in the tunnel's sidewalls; foul air is extracted through openings in the ceiling. The transverse system is suitable for long tunnels that can be divided into a series of air circulation cells, so that foul air does not have to be transported the full length of the tunnel to be vented. Many tunnels—including the Chunnel—use a combination of transverse and longitudinal systems. The Chunnel has a fan station at the English and French coasts from which air will be supplied to the central service tunnel, and distributed from there to the running tunnels on either side. There will also be two supplementary fan stations for direct supply and extraction of air in the running tunnels.

The 100-foot-deep Peachtree Center Station, shown at right, is part of the expansive Metropolitan Atlanta Rapid Transit Authority (MARTA)—the modern subway system beneath the streets of Georgia's capital city. Although most of the tunnels were bored through soft ground, the 60-foot high, 42-foot-wide arched station was excavated by blasting in solid granite. One hundred and fifteen thousand people will pass through the station daily by 1995. The Peachtree Center Station and the network of tunnels that connects it with the many others in the MARTA system are serviced with an advanced ventilation system that not only brings fresh air to the users, but offsets the heat generated by the trains as they hurtle along the tracks.

The need for subways, and tunnels of all types, will not diminish in the future. Urban overcrowding, traffic congestion and the need for more efficient water and sewage systems will demand the construction of more tunnels. Urban subway systems may be bored deeper beneath the ground, perhaps through hard rock strata. New tunnel boring machines are able to deal with increasingly harder rock conditions. Soft ground TBMs are also becoming more advanced, and soon the erection of tunnel liner segments, which still involves a good deal of manual labor to install them, may become completely automated. Geological prediction and evaluation of ground type at the tunnel site are considered high priority areas of research and development, and may be spurred on and enhanced by the use of computers. With the success of the Chunnel, other long and ambitious subaqueous tunnels will be planned and constructed. One idea, first suggested almost a century ago, is to build a tunnel under Bering Strait that would connect the Soviet Union to mainland United States. The tunnel, joining Russian Siberia to Alaska, would be about 30 to 40 miles in length. If this hypothetical tunnel is ever built, a person would be able to drive from New York City to Alaska, across the Soviet Union, and through Europe, take the Chunnel under the English Channel and finally arrive in London, England.

INNER SPACES

I
n ancient Rome, 50,000 spectactors at a time crowded the enormous Colosseum to indulge their taste for gladiatorial combat, lounging comfortably in the shade of a gigantic awning stretched across the arena. On today's sporting scene, the gladiators live to perform another day, and the spectators enjoy ever larger and more luxurious covered stadiums with as many as 90,000 people gathering beneath huge spanning roofs that shield them from the elements. The impulse to assemble in large numbers is played out in other vast inner spaces as well: Domed temples, cathedrals and mosques inspire worshippers and rise as landmarks in a city's skyline; concrete spans arch over throngs of travelers at airport terminals; and tents of awesome dimensions afford protection to tens of thousands of spectators at various events.

In size, shape and materials, modern wide-spanning roofs far outstrip those that predated the 20th Century. The widest of them span more than 600 feet; the shapes range from traditional hemispherical domes to whale-like humpbacked structures; and instead of brick and stone, engineers and architects now build with steel, reinforced concrete and paper-thin acrylics. Yet the great spans of the past and the present share the mere handful of physical principles that prevent them from collapsing under their own weights and other loads. These structures fall into one of two groups—compression or tension structures—according to the kinds of stresses they must endure.

The first group includes the classic domes made of stone, brick or concrete. From the outside, these compressive domes look like a sphere cut in half to form a hemisphere. Working on the same principle as the arch, these spans channel a load downward toward their bases. The challenge in such large roofs always has been to prevent bending under their great unsupported weights, and to corral the spread, or end thrust, that occurs at the bottom of all arches. The Greeks, who built stone temples such as the famed Parthenon atop the Acropolis in Athens,

---

The 10-acre, acrylic roof of the SilverDome in Pontiac, Michigan, covers a huge stadium that seats over 80-thousand spectators. This enormous membrane roof is given strength and rigidity by a lattice of steel cables. Although it weighs 200 tons, the roof is supported only by the tension of the air pressure created by 15 electric fans.

tried to emulate wooden post and lintel structures, but never solved the bending problem in their large edifices. It was the Romans who so effectively exploited the strength of the arch and used it to create the basis for the wide-spanning roofs. Little changed for nearly two millennia, until the arrival of reinforced concrete late in the 19th Century. The new material opened up a whole genre of long-span, lightweight roofs that cover vast inner spaces, relying on the tensile strength of steel and compressive strength of concrete. These "thin shell" roofs, which were significantly lighter than traditional domes, took on shapes and sizes that would have startled Roman builders.

Other kinds of spans covering huge inner spaces are primarily tension structures. Some work very much like a suspension bridge, relying on steel cables—sometimes in conjunction with concrete—to support wide roofs. One type, founded on the timeless principles of the tent, is the model for another modern tension span; but here acrylic membranes replace the skins and steel cables replace the ropes. A third kind of tension structure resembles a bubble and relies on internal air pressure for support. These inflatable domes provide relatively inexpensive roofs that cover everything from tennis courts and long, wide playing fields to huge auditoriums and circuses.

ANCIENT SOLUTIONS

One of the persistent major goals of architecture has been to enclose a roomy, open area, unencumbered by walls or columns. For more than two millennia, the best solution was to raise compressive masonry or concrete domes. However, large domes, like most wide-spanning structures with no direct support from below, tend to bend under their own massive weights, and can deform and in the worse case eventually collapse. Therefore, the builders of domes have been on a quest to find the right materials and techniques to reduce tension, while still providing the desired large roof span overhead and unobstructed inner space. Until the 20th Century there were traditionally two solutions. The first was to build a classic compressive dome. The second was to build a structure that looked like a traditional dome, but actually worked on different principles.

The St. Paul's Solution
St. Paul's Cathedral in London, England,is a three-layered dome *(outlined in purple.)* The outer dome is made of lead-covered wood and supported by an internal wooden framework. An intermediate brick cone supports the framework and the huge lantern above the dome on top of the cathedral. The inner dome, which is seen only from the inside of the building, is made of brick. Because it is relatively lightweight, the inner dome requires less support around its perimeter.

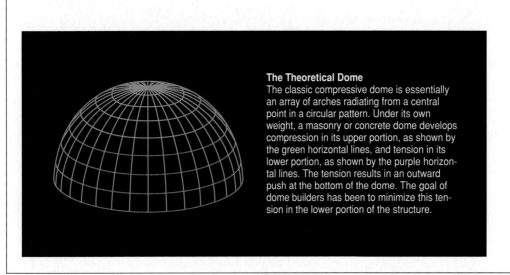

The Theoretical Dome
The classic compressive dome is essentially an array of arches radiating from a central point in a circular pattern. Under its own weight, a masonry or concrete dome develops compression in its upper portion, as shown by the green horizontal lines, and tension in its lower portion, as shown by the purple horizontal lines. The tension results in an outward push at the bottom of the dome. The goal of dome builders has been to minimize this tension in the lower portion of the structure.

CLASSIC DOMES

For centuries the traditional dome was the best way to roof over wide spaces that might extend to 100 feet or more in diameter. Like the arch, the dome tends to channel a load along its axis so that it follows a curving path through the structure. In fact, the dome acts like a series of theoretical arches all meeting at their crowns over a central point. Moreover, the dome garners support from the other arches allied with it, and so is stiffer—and can be built far thinner—than a single arch of the same radius.

A model of a traditional dome looks like a peeled orange cut in half perpendicularly to its segments. Each section is the equivalent of a half arch; the theoretical divisions between them are called meridians. The arches lean against, and support each other, and each is essential for the others' stability; if one is plucked out, the dome collapses. The dead load of the dome tends to flatten its crown slightly, compressing the higher portions of the arches. In the lower half of the dome, however, the dome pushes outward, its end thrust stretching the material

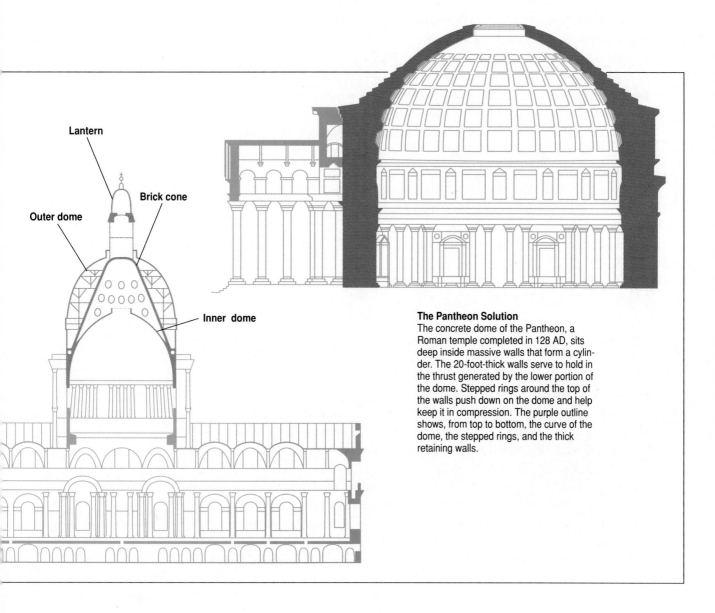

The Pantheon Solution
The concrete dome of the Pantheon, a Roman temple completed in 128 AD, sits deep inside massive walls that form a cylinder. The 20-foot-thick walls serve to hold in the thrust generated by the lower portion of the dome. Stepped rings around the top of the walls push down on the dome and help keep it in compression. The purple outline shows, from top to bottom, the curve of the dome, the stepped rings, and the thick retaining walls.

Labels: Lantern, Brick cone, Outer dome, Inner dome

in tension. Therefore it is important that this tension is withstood by a series of abutments or a girdling tension ring at the bottom of the dome.

The effort to control this outward thrust inspired the brilliant design of the Pantheon, the famed Roman temple completed in 128 A.D. One of the best-preserved large buildings of the classical era, it is also one of the world's most copied structures. The structure of the Pantheon is essentially a thick-walled cylinder on which sits a 142-foot-wide hemispherical dome—a diameter audaciously daring for the time. In fact, no other dome would equal the Pantheon's diameter for 13 centuries. From the outside, the dome looks relatively low. Only from the inside of the temple is the complete area of the hemisphere, which extends down within the cylindrical wall, completely visible. The distance from the floor of the temple to the crown of the dome also measures 142 feet, giving the entire structure the visual symmetry of a perfect sphere. The only window is a brick-lined oculus—a round, unglazed 26-foot opening in the top of the dome.

The dome, which is 5 feet thick at the top, gradually thickens to 21 feet at the point where it meets the stout walls. Just above this level, a series of stepped rings of brick rest atop the concrete walls and help keep the dome in compression. The dome's base sits well down inside brick walls which serve to counteract the dome's tendency to push out. A modern inspection of the Pantheon's dome revealed numerous—but apparently harmless—cracks in the concrete in the lower portions. In an effort to analyze how effectively the stepped rings and supporting wall work, subsequent computer modeling showed that the Pantheon does indeed act as the theoretical model dictates; most of the cracks occur in a radial pattern near the dome's lower boundary where tension occurs. Further modeling showed that a hypothetical Pantheon, with the stepped rings removed, would be much more heavily cracked. The models show how effectively the thick walls and rings resist the tensile forces by keeping the concrete dome in compression.

Although Roman engineers had already mastered the technique of building smaller domes for temples and bath houses, the scale of the Pantheon's dome was without precedent. The choice of concrete for constructing such a monumental building was another departure from tradition. Concrete offered practical advantages over brick and stone, the traditional materials for temples, palaces and other important buildings. Long restricted to utilitarian roles such as foundations, concrete was plentiful and far cheaper than cut stone or brick, which required the services of skilled masons. Construction also moved at a much faster pace with concrete, and inexperienced workers could learn quickly how to handle it. Unlike modern concrete, the Roman version was so thick that it was applied by hand in layers instead of being poured into molds. Aside from its economy and ease of handling, concrete proved to have other properties that made it an ideal material for a large span. Of particular importance was that it could be applied in whatever thickness was needed at a given point.

For many hundreds of years after the Pantheon was completed, the physical form of the dome—a hemisphere—remained a popular shape for roofs over large inner spaces. Unlike the designers of the Pantheon, later builders attempted to expose as much of the dome as possible to outside and inside views. However, because of the weight of masonry and concrete, and the problem of tension, other designs were developed to achieve the same visual effect. When the noted English

Frozen Domes

The Eskimo igloo is a dome shape made of blocks of snow that may be laid in a continuous spiral or concentric rings that taper toward a flattened top. The builder sometimes fits the clearest piece of ice available into the upper opening so that light can pass through. Loose snow is packed into the chinks, the entrance is blocked, and a fire is lit inside. As the walls soften, water trickles down the curving contours. When the entire inner surface is damp, the fire is put out and the entrance opened. The blast of frigid air hardens the joints, providing additional stiffness and an inner surface that acts as a consistent barrier against the frigid winds of the Arctic.

An Eskimo builder fits one of the last snow blocks in the making of an igloo. The completed hemispherical igloo is aerodynamically efficient in the face of strong northern winds; as a dome it is structurally strong.

architect Sir Christopher Wren designed a dome for St. Paul's Cathedral in London in the late 17th Century, he understood well the relationship between the shape of a dome and the way it worked—in particular he appreciated that the classic dome shape need not work in the traditional way: a single, thick, monolithic structure bearing its own entire weight. Wren was a man of many talents— astronomer, physiologist, mathematician, inventor—and was England's preeminent architect. His St. Paul's achievement may have been founded on the failures of earlier dome builders. Wren probably was aware that the dome of St. Peter's Cathedral in Rome was afflicted with problems. Although it had been designed by none other than Michelangelo, and was inspired by the Pantheon, it was plagued with cracking after its completion in 1593. The dome's supporting walls were not thick enough to contain the outward thrust in its lower portions. By Wren's time, chains of wrought iron bars had been wrapped around the base of the St. Peter's dome, but they compensated only partially for the structural errors.

The dome of St. Paul's, which looks like a classic hemispherical structure from both outside and in, is in fact a three-layered structure—one dome visible from the outside, another dome visible from the inside and a brick cone, sandwiched out of sight between the two. The three elements work together to create a large and beautiful structure that has endured, among other things, the London blitz during World War II. The outer dome of St. Paul's is made not of masonry, but of timber sheathed with lead; it is light and serves a purely aesthetic function. The innermost dome, which is made of brick and spans 100 feet, forms the cathedral's

HARD-TOP CONVERTIBLE

Completed in 1989, the SkyDome in Toronto, Canada, is the first dome to feature a fully retractable, hard-shell roof. The steel dome has a span of 682 feet and is 282-feet high. The inset illustration shows how the dome's roof opens. The green segment remains stationary at all times. The yellow segment travels on a circular track along the outside perimeter of the dome and stops above the green segment. Finally, the pink and purple segments move along trolleys in a line straight back to cover, one atop the other, the yellow and green segments. The entire process takes 20 minutes.

ceiling. Hidden between the two is a tall brick cone with walls 18 inches thick. The dead load of the outer dome and the live loads it bears, such as wind, are transmitted to the cone through a timber framework. The cone in turn transmits the loads into the supporting walls. In addition to the outer dome, the cone supports a towering stone lantern atop the outer dome. Its 850-ton weight helps to compress and stabilizes the cone, which in turn contributes to keeping the inner dome in compression. Around the cone's base Wren placed a chain, which computer analysis now shows is extremely effective against outward thrust. St. Paul's dome has been the model for many other so-called false domes, including the U.S. Capitol Dome in Washington, D.C., completed in 1864.

The classic dome shape and principles remain in use today for numerous churches and public buildings. But in many cases, the form is all that remains; sophisticated materials have allowed entirely new principles to be applied for the same effect. The gargantuan Louisiana Superdome in New Orleans is one of the world's largest domes, with a diameter of 680 feet and a towering height of 273 feet at its center. The stadium can hold 75,000 people for sports events and 20,000 more for rock concerts and gatherings such as political conventions. The dome-like shape is achieved by a steel framework that has been covered with a non-weight-bearing "skin." The huge, slightly flattened roof's major structural component is a skeleton composed of 12 arching steel ribs radiating from the crown, further connected by six circumferential horizontal steel rings. The vast frame is covered with a three-layered membrane of sheet steel panels topped with polyurethane foam and then a sprayed-on coat of weatherproof plastic. And although the dome is made of steel, which withstands the massive tension well, tremendous lateral thrust is still created at its base. The outward push is controlled in a traditional way, by means of an enormous steel tension ring.

The use of steel ribs for some modern domes has allowed one important element of classic domes to fall by the wayside. Traditionally, a dome's stiffness—and integrity—depended upon a constant distribution of forces through a solid structure. If a section of it were removed, collapse would follow. Some modern domes, particularly in astronomical observatories, can still work with a portion removed. Recently, this has been taken to its ultimate conclusion. Despite its name and its dome-like profile, Toronto's 682-foot-wide, steel-covered SkyDome is structurally unlike a true dome. The roof of the Skydome is composed of four steel sections that are structurally independent, but which work together, not unlike a set of graduated mixing bowls that fit within each other. At its north end is a fixed section, over which the other three sections—each, in effect, a single slice of a proper dome—slide on trolleys along tracks, one atop the other, to open the stadium roof in good weather. The dome will not collapse because each part is entirely self-supporting. Yet, like true domes, the SkyDome is also a compression structure, the full 9,500 tons of its retractable roof being channelled downward into the thick walls of the stadium.

THIN SHELLS

One of the most significant developments in the history of huge roofs over inner spaces occurred during the late 19th Century. After centuries of structural hegemony, stone and brick were displaced in a revolutionary and wholehearted return

The House With a Troubled Past

The winning design in a 1957 architectural competition for an opera house in Sydney, Australia, was nothing less than inspired—a building with seven distinctive roofs, similar in shape to the wind-filled sails of the boats that would glide past on the waters of the harbor. Architect Jørn Utzon's vision is perhaps the best-known example of a design that worked on paper, but was incredibly difficult to translate into reality.

Utzon planned the 200-foot tall roofs as a cluster of self-supporting thin shells of concrete. But the consulting engineers realized that shells of that shape on such a large scale would not stand up under their own weight nor in the face of gusting winds. Rather than alter Utzon's original vision, engineers sought a successful way to construct the same form. Instead of a thin shell cast on site over formwork, huge precast concrete ribs were constructed. These ribs, each weighing about 10 tons, were paired to create forms that had been utilized since ancient times: arches. The arches, placed side by side, fanned out from the base and were joined at the top by a concrete beam. The resulting framework was covered with precast concrete panels and ceramic tiles.

The ingenious technique worked, and although it was far behind schedule and very much over budget, the opera house, completed in 1973, was faithful to Jørn Utzon's dramatic vision. It very quickly became not only a national symbol, but also a monument to the ingenuity and capabilities of the project's structural engineers.

Sunset gilds the roofs of the Sydney Opera House, seen in front of the Sydney Harbour Bridge. During construction (inset), the strengthening ribs, connected to form pointed arches, were visible.

to concrete. The reason for concrete's revival as a material suitable for wide spans was the introduction of metal reinforcement to give it tensile strength. The Pantheon, for almost two thousand years the record holder for concrete domes, was demoted to second place in 1913 by a reinforced concrete dome in Breslau, Germany, whose inside diameter measured 216 feet.

Although its material was revolutionary, the Breslau dome, like the Pantheon, is a compression structure that acts like classic domes: It is in compression in the upper portions and tension in the lower portions. Although the dome was not nearly as thick as the Pantheon, German engineers—who were pioneers in the field of reinforced concrete roofs—quickly discovered that it could have been much thinner still. Not nearly as much material was needed to guarantee the stiffness of a dome made of reinforced concrete. New domes were pared down in thickness, and in the process became what is now known as the "thin shell." The Market Hall, completed in 1929 in Leipzig, Germany, had two domes. Each spanned 247 feet, but was only one-third the weight of the smaller Breslau Dome.

As a consequence of its remarkable thinness, the shell is lighter, does not develop excessive tension in its lower portions and does not need a heavy girdling ring or thick supporting walls. However, although the shell must be thick enough to carry loads—primarily its own weight—it must not be so thin that it cannot withstand loads at precise points, or a puncture. This reflects the physics of an egg, which can withstand extreme compressive pressure throughout its form, but not a sharp load applied to a small specific area. The construction of thin shells presents engineering and financial challenges as well. The concrete for these structures must be poured into carefully designed formworks. It can be difficult and dangerous work to build up the scaffolding and forms at great heights. Moreover, the thin-shelled roofs often assume complex curving shapes that may require numerous individual form shapes to be designed, manufactured and placed, before pouring can take place. The formwork must withstand the weight of the concrete, and so is a major structure itself.

Shape, which creates stiffness, is another of the thin shell's determining characteristics. When a flat piece of paper is held between between a thumb and finger, the paper droops. But when that piece of paper is bent slightly along its length or width, it becomes more rigid. Whether it is a sheet of paper or a building, any object stiffened by curvature is called form-resistant and may take on a variety of shapes. The three principle configurations are the hemispherical dome, the saddleback, which is curved along both its length and width, and the barrel vault, which is essentially a half-cylinder. The dome and the saddleback are stiffest because they are curved in two directions and are defined as non-developable; that is, it is theoretically impossible to flatten either shape without deforming the material. A barrel vault, on the other hand, which has the shape of a classic arched airplane hangar, is not so stiff because it is curved in only one direction and is therefore developable; in theory it could be flattened into a rectangle without any deformation of the material.

The combination of fewer materials and stiffness through shape works to make thin shells light but strong. The thin shell dome also from benefits from the strength and stiffness inherent in the classic shape, making it the most efficient of shells. St. Peter's, the famed masonry dome in Rome, has a span-to-thickness ratio of 1 to 14; an egg ranges from 1 to 30 to 1 to 50. In a reinforced concrete shell the ratio is on the order of 1 to 400 or more. In practical terms, this means that a reinforced concrete dome, just 3 inches thick, can span a minimum of 100 feet. Yet a thin shell dome still undergoes some tension in its lower portions and is usually reinforced with a tension ring. The modern hemispherical dome, like its models from the past, typically spans public spaces such as planetariums, churches, temples and mosques. The Jefferson Memorial, in Washington D.C., pays homage to its roots with a series of stepped rings like the Pantheon's. The Memorial's rings, however, are for purely visual effect; metal reinforcement shores up the dome's vulnerable lower boundary.

Most domed thin shells have a shallow, flattened curve, which offers advantages of economy over a full hemisphere. By reducing height, the total volume of the roof is also reduced, but without sacrificing any floor area or usable space. Flattening the curve also reduces the weight and the resulting dead load. The trade-off of flattening a thin-shelled dome, however, is increased outward thrust at the base. In the 197-foot domed Palazetto dello Sport, designed for the 1960 Olympics in Rome, Italian engineer Pier Luigi Nervi marshaled several techniques to withstand this stress. Taking advantage of form resistance—strength through curvature—he molded thickening scallops into the outer edge of the dome, which then behaved as if it had been thickened. To gain further support, Nervi braced the Palazetto around the perimeter of the roof with Y-shaped, concrete buttresses, or supports. These flying buttresses serve as a continuation of the dome's shallow curve, transmitting the load to an eight-foot-wide tension ring that also works as

Shells In Nature

Many roofs directly reflect naturally occurring shapes. Thin-shelled structures, for instance, resemble eggs, which are remarkable in their ability to withstand even pressure from top to bottom, but which will puncture when pressed sharply at one point. The rippling or scalloping of some roofs reflects the scallop shell, which is fluted—in effect thickened—for additional strength. And a walnut half shell is similar to a heavy compressive dome; its thickened lower edge works against outward thrust in a similar fashion to the way a tension ring works around the lower perimeter of a dome.

the building's foundation. The underside of the roof is stiffened by the addition of a lattice of intersecting ribs. While this presents a distinctive diamond-shaped pattern of visual interest, its real purpose is to help prevent buckling.

However, several Palazetto domes would fit comfortably beneath the expansive roof of the Centre National des Industries et des Techniques (CNIT), a thin-shelled exhibition hall in Paris. Completed in 1958, the 676-foot span consists of a double-layered reinforced concrete shell, composed of three elongated sections which join to give the impression of a dome, but with a difference. It is as if three large portions from the sides of a hemispherical dome had been cut away, leaving the roof balanced on its three remaining corners and producing a triangular floor plan. The problem with huge shells such as the CNIT is not the overall stress they must endure throughout their forms, but the possibility of buckling. This sudden and unpredictable situation can result from very specific loads—known as point loads—applied to a limited part of the shell. Because of their extreme thinness, the shells are extremely vulnerable to loads produced by anything other than their own weight and weather. All thin domes, then, must be stiffened. In the CNIT, sandwiched between the two 2 1/2 inch-thick layers of the shell, is a series of stiffening vertical ribs or diaphragms. In addition, the outside of the roof is corrugated, or rippled, the curvature serving to imitate thickening and making the entire shell even more rigid.

The 676-foot-wide roof of the Centre National des Industries et des Techniques (CNIT) in Paris, France, is one of the widest thin-shelled structures in the world. Although dome-like in appearance, the roof actually is fashioned from three sections that meet at the center. The span is made of two nearly identical layers, or reinforced concrete shells, each only 2 1/2 inches thick. The structure's designers stiffened the roof by corrugating the surface and creating a lattice of ribs between the two layers.

GRACEFUL SADDLE SHAPES

While it is stability and symmetry that characterize thin shell domes, the sweeping curves of a saddle-shaped structure—another thin shell—look like motion arrested. Here, a longitudinal curve sweeps down, as if from the front of a huge saddle, running along its length, and then curving up at the saddle's rear. At right angles is the second curve, which arches down. Because of the unique configurations of its curves, the saddle shell's technical name is a hyperbolic paraboloid, or hypar for short. It is as if a four-sided frame had been twisted or warped to produce a double curvature extending in opposing directions. This warping makes the saddle shape nondevelopable and therefore inherently stiff. The downward, side-to-side curve acts like a series of compressive arches, while the longitudinal curve mimics the sweep of a suspended cable in tension, exerting an inward pull on the two end points of the saddle. The opposing double curvature gives a concrete shell great rigidity.

A great master of the concrete hypar was architect and structural engineer Felix Candela, a Spaniard who emigrated to Mexico when General Franco seized power in Spain in 1939. In his new country Candela built numerous restaurants, churches and industrial buildings with dramatically curved roofs. His first hypar, the Cosmic Ray Pavilion in Mexico City, still holds the distinction of being the world's thinnest concrete roof. The laboratory's shell measures a scant .59 inch in thickness and spans 35 feet. Candela followed his first effort with more hypars, both single and multiple. For a nightclub 140 feet in diameter, eight hypars radiating from a common center form a distinctive undulating roof that extends downward to the ground. In effect, the roof is the entire structure. Joined edge to edge, the ends of the hypars create a perimeter that gives the appearance of eight rounded peaks arranged in a circle and separated by eight equally sized troughs; the bottoms of the troughs reach the ground and serve as the supports of the building. The load, which is relatively light since the concrete is 1.5 inches thick, flows from the roof—the upper surface of the hypars—through the troughs where it is supported by the ground. The deep undulations also serve to stiffen the nightclub's shell, in exactly the same way the scalloped rim of the Palazetto dello Sport does, but on a much larger scale.

Despite its double-curved complexity, the geometry of the hypar actually makes it easier to construct than a classic dome-shaped thin shell. In the dome, all the lines are curved, so the entire formwork required for pouring the concrete must be carefully constructed from curving pieces. The hypar, however, has what is called a rule surface. From any point on it, there are two directions in which the surface is straight. Building the formwork for casting the concrete is therefore less expensive because ordinary straight pieces of timber can be used.

A third thin shell structure looks like a classic Roman barrel vault and acts as if a series of arches were joined together to form one elongated arch. In the traditional masonry barrel vaults, tension develops in the bottom portions where the outward thrust is strong. A traditional barrel vault requires supports or buttressing to withstand the outward push along its entire length. However, if the vault rests not on the ground or along a solid wall, but is held up only at each end, it will, in theory, act like a bending beam—its top in compression and its unsupported lower portion in tension. However, standard concrete or masonry cannot withstand the bending, so reinforced concrete is used. With such stiffening, vaults up to 40 feet long, and with extremely shallow curvatures, can safely extend—unsupported at one end—over a building's entrance. It is the curvature—in this case identical to the curved piece of paper—that assures stiffness. Arrays of thin-shelled vaults connected side-to-side have been used to extend over the seating areas of outdoor arenas, providing shelter for spectators. Such arrays of barrel vaults are used to make roofs for factories, airplane hangars and other such utilitarian buildings.

GEOMETRIC LATTICE

The spherical geodesic dome, made famous by the flamboyant promoter-designer Buckminster Fuller, works like a thin shell, even though it is not made of reinforced concrete. The geodesic dome is divided into a lattice of geometric units—usually triangles, pentagons or hexagons. The units combine to form a structure that is almost an entire sphere. The smaller the dimensions of the units, the more like a thin shell the dome becomes because the entire surface—the metal borders and acrylic—act as a single material in the same way concrete and steel act in unison in reinforced concrete. In fact, all of the units work together to form a monolithic structure that functions like a thin shell: It is light and gains strength from curvature. Usually constructed of aluminum or steel bars or pipes and acrylic plexiglass or aluminum plating, the geodesic dome is extremely light compared to solid domes of the same dimensions, sometimes weighing only 1/300th as much, and requiring only a modest foundation. Construction of a geodesic dome can be simple: The lightweight materials that are commonly used are easy to work with, and the structure is composed of relatively few types of components that can be mass-produced and quickly assembled.

Walter Bauersfeld invented the geodesic dome in 1922. The German engineer was one of a small group of men who not only had designed new thin shell domes made of reinforced concrete, but who also sought to understand the mathematical basis for them. The highly geometric geodesic dome was probably a byproduct of Bauersfeld's investigations. Nevertheless, it took the enthusiasm of Fuller to popularize it. He demonstrated its practicality for inexpensive, quickly assembled housing, restaurants and other small-scale uses. But the new dome failed to make a big splash until 1958, when a hemisphere 384 feet in diameter was erected in Baton Rouge, Louisiana, to house a repair shop for railroad cars. Since then, the geodesic dome has been adapted for other industrial purposes, museums, auditoriums and athletic arenas. For the American pavilion at Expo '67 in Montreal, Canada, Fuller designed a 250-foot dome that was virtually a complete sphere; the first fully spherical geodesic dome, however, is the Spaceship Earth, part of the Epcot Center at the Walt Disney World theme park in Florida. Looking like a gigantic golfball propped up on six massive steel legs, the 165-foot sphere has a steel framework covered with reflective aluminum panels. Grand though they are, none of these geodesic domes approaches the colossus Fuller dreamed of—a structure two miles across covering part of Manhattan in which an optimum environment could be created. According to Fuller, even a dome of such dimensions could be completed in three months, with helicopters ferrying workers and materials to the structure's upper reaches. However, geodesic domes—despite the fact that they have been around for many decades—have never been implemented

The Spaceship Earth at the Epcot Center in Walt Disney World near Orlando, Florida, is the first fully spherical large geodesic dome in the world. With a diameter of 165 feet, the dome is made up of nearly 1,000 triangular aluminum panels bolted to a spherical steel framework, and covered with waterproof sheeting. Standing on its six steel legs, the dome is 180 feet tall. The opaque outer skin means that, for the visitor inside the building, the impression is of infinite space.

on a widespread, practical scale. The distinctive domes are most often used as pavilions and greenhouses, or as alternative housing. But geodesic domes have never lived up to their promise.

SPANNING THROUGH TENSION

Classic domes and modern thin shells are all designed to minimize the bending that can be the downfall of any wide-spanning roof. However, there is a wide range of large roofs and stadiums that not only endure tension, but also rely on that same tension for their very structural integrity. Like thin shells, these roofs often depend on the union of steel cable and concrete working together as a single material; in many other cases entirely new systems and materials have provided the basis for these tensile roofs.

Over the last decade, a new type of large hemispherical roof—the pneumatic dome—has become a familiar form on a growing number of landscapes. Made of strong fabric and a network of wire cables, these air-inflated domes are the exact opposite of compression domes; every inch of the inflated surface is in tension. The great roof of the 61,500-seat, 193-foot-high, Hoosier Dome in Indianapolis, Indiana, is one of the best known of the pneumatic structures. Completed in 1983, the roof of the arena is made up of a vast fabric membrane, with a total area of eight acres and a weight of 257 tons. Yet it is supported by air pressure alone. To supplement normal air pressure, compressed air at five pounds per square foot is blown against the roof's inner surface by 20 electric fans. The material roof has two layers, each of them measuring an amazingly thin 1/32nd of an inch thick. The weatherproof outer layer is Teflon-coated fiberglass. Fiberglass has an extremely high tensile strength, but it weakens when wet—a problem controlled by adding Teflon, which shields the fiberglass against moisture, as well as potentially damaging high temperatures and ultraviolet rays. Because fiberglass reflects sound-waves and makes for a noisy stadium, sound-absorbing canvas is used to line the inner surface.

Both layers of the Hoosier Dome roof are considered membranes—structures so thin that the only load-bearing force that develops in them is tension. How large a load a membrane can carry is determined by its tensile strength, which in turn is affected by how steep or shallow the curve is. A flattened curvature requires less tension, or air pressure, to support the membrane. It also is a more economically efficient shape, requiring less material to cover the same area than is required for a higher dome. However, a flattened dome will also be subject to additional loads, such as snow, which could bring on deflation or collapse. One pneumatic dome—the MetroDome Stadium in Minneapolis, Minnesota—did, in fact, deflate because of snow loading. The Hoosier Dome membrane, however, is further reinforced by eight intersecting steel cables spanning its central portion; strung out, the 3 1/2-inch-diameter cable would stretch two miles. At each end the cables are anchored to a reinforced concrete ring that circles the bottom edge of the roof and keeps them in place. The cables help to resist the upward force of the internal air pressure by strengthening and stiffening the fabric roof.

Although the idea of simply inflating a membrane roof seems, and is, a simple idea, a great deal more technology is required to keep it working. At the Hoosier Dome, roof conditions are constantly monitored by a weather and computer station that can detect flutter or an excessive buildup of snow. Special measures systems can be installed to compensate for this additional load; hot air is pumped through the space separating the two layers of the roof to keep it free of snow.

SUSPENSION ROOFS

Inflating a membrane like a balloon is not the only way to develop tension in a wide-spanning roof. Suspending it is also a common technique; one type of suspension roof works in essentially the same way as a suspension bridge. High overhead cables, which may be strung between towers, arches or the walls of a building, can support a roof, which then needs no columns or other members cluttering the inner space beneath. However, a suspended roof, like a bridge, must deal with the problem of stiffness and susceptibility to wind; the problem is flutter, or the tendency to flap and oscillate like the large sail of a stalled boat in even moderate winds. Therefore structural measures must be taken to reduce this dangerous fluttering. In one system the cables may be permanently tensed, or prestressed. In other cases the concrete that encases the cables will be rigid enough to minimize any flexibility. Or the roof may be secured and tensed by cables and guy-wires that extend all the way to the ground.

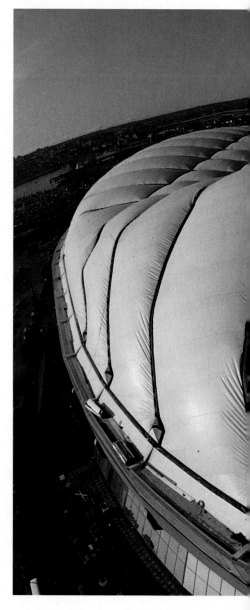

The fabric and cable roof of the British Columbia Place Stadium in Vancouver slowly inflates to its full height of 200 feet on a cushion of air from 16 electric fans. The 430,000 square feet of material used for the dome is a double-layered, fiberglass-woven fabric coated with Teflon. Incredibly, each layer is only 1/30th of an inch thick. The 22 steel cables gives the inflated dome extra rigidity and strength. Completed in 1983, the dome covers an inner space of 760 feet by 623 feet.

With its system of towers and cables, such roofs can span great distances—more than 800 feet in some structures. And the roof, which is normally composed of reinforced concrete slabs, is totally independent of the building's walls. The entire roof is in tension; there are no trusses or girders that bend.

The horizontal suspension-bridge roof is an economical way to enclose huge unobstructed spaces such as factories, exhibit halls and convention centers. But there are a wide variety of shapes and configurations that can be applied to roofs, all of which exploit the excellent spanning strength of high tensile steel cable. The J.S. Dorton Arena, a livestock arena in Raleigh, North Carolina—completed in 1952 and nicknamed the Cow Palace—has a radically innovative support system for its network of cables. Two enormous reinforced concrete arches lean toward each other, eventually crossing near their bases. A grid of cables is strung between the arches providing the support system for the roof. The system has been compared to a director's collapsible wood and canvas chair: when opened up, two

The Original Mobile Home

In the mid-1980s, climbers advancing on the high and wind-swept slopes of Mount Everest spotted a splash of color peeking through the surface of the blinding snow. After some digging, the mountaineers uncovered a thin nylon tent that had been abandoned by a team the year before. Amazingly, after 12 months of pummeling by Himalayan storms, the tent was still intact, a testament to the strength and efficiency of these remarkable tension structures.

Tents—from the tepees of plains Indians, to the felt yurts of the Mongols have been in use for centuries. Their trademarks—extreme light weight, an unobstructed inner space, the ease with which they can be transported and assembled and their tremendous strength in the face of sometimes hostile natural forces—have not changed. But the last 30 years have seen a revolution in designs and materials; tents are now lighter, stronger, and quicker to erect.

Waterproofed nylon tents were first used during World War II, but the new fabric did not "breathe" and water or ice from condensation built up inside. The solution, developed during the 1960s, was a breathable nylon inner shell protected by a suspended, waterproof "fly" on the outside. Aluminum replaced wood or steel for poles, reducing weight without sacrificing rigidity. In the early 1970s, innovative geodesic dome tents were developed. Here, four or six segmented aluminum "wands" are inserted into long matching sleeves on the outside of the fabric walls; the flexible poles are bent into thin transecting arches and slotted into rings at the outside corners of the tent floor. The arches combine to form an external skeleton that puts the nylon into tension, creating a surprisingly rigid and strong domed structure that can be erected in minutes. Some domed tents can withstand a load of 250 pounds on the crown. Members of one expedition to the Arctic's rugged Ellesmere Island reported that their tents easily withstood consistent 60-mile-per-hour winds. And the Everest experience proved that extreme snow loads are no problem.

Now there are countless varieties of tents, starting with the traditional A-frame and including many permutations of the dome. All are made of fire-retardant materials; many are reinforced to withstand the elements during three or four seasons. Good tents will be cut to be uniformly taut, helping them to deflect winds. They are characteristically light as well; among today's hikers and climbers, if a two-person tent weighs 10 pounds it is considered to be heavy.

The domed tent in this photo was used during a 1977 Ellesmere Island expedition in the Arctic and withstood icy winds of 60 to 70 miles per hour.

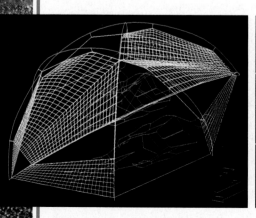

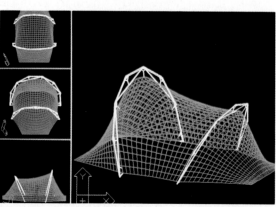

Today's tent designs, such as the ones at left, are made with CAD, or computer-aided design. Far left is the design for a geodesic dome that will hold three people. Four transecting aluminum wands keep the tent taut in tension. The other is a hoop, or tunnel, tent that is much lighter than the dome, but still strong against the elements. It relies on two light compressive aluminum arches for support.

hinged struts cross each other, but are prevented from collapsing by the canvas seat that runs between them. The combination of the tensed canvas and the rigid, crossed struts gives the chair stability. In the Cow Palace the struts are the crossed arches, while the canvas sheet is the roof. For further reinforcement in the structure, steel columns around the outside help support the arches and provide the frame for the outside walls. The cables running the length of the roof have a concave sag, while those in the opposite direction are convex. The result is similar in form to the thin shell, saddle-shaped roof; however, the entire system of cables is in tension. The grid of cables supports lightweight metal roofing panels, which are stabilized against flutter by internal guy wires. This ingenious system for a large-span roof inspired other variations of the suspended cable network. A similar vintage hockey rink at Yale University has a wood-covered suspension roof that looks like a humpbacked whale. The main structural member is a single, spine-like arch of more than 300 feet running the length of the arena. Perpendicular cables extend outward from both sides of this arch and are anchored to the long, exterior walls that run parallel to the spine. A second set of cables runs the length of the arena, crossing the first set at right angels. This creates the web for the two sloping sides of the roof. The combined weight of the wooden covering between the cables and some internal trussing stabilizes the roof. Although the hockey rink and the Cow Palace have numerous individual cables in tension, in both buildings the roofs act as if they were continuous membranes like the material stretched over the Hoosier Dome.

A particularly beautiful roof that relies on suspended cables spans 214 feet over the Dulles Airport passenger terminal near Washington, D.C. Here, though, the strength of the roof results from concrete being poured around the network of cables to create a rigid span that works like reinforced concrete. Nevertheless, the monumental roof has a catenary curve—the shape that a suspended cable takes when the only load it bears is its own weight. This is a deceptive image, however, because the cables actually bear much more than their own weight. At Dulles, 16 parallel cables are strung between two rows of concrete pylons, one 48 feet tall and the other 65 feet tall; each row is angled away from the other to counteract the inward pull of the cables. The cables are kept tensed by the weight of the roof's concrete panels, which were poured in place around them. To support the enormous weight of the roof, the pylons are heavily reinforced with steel rods—each of the taller pylons contains 20 tons of steel.

Suspended cables are also at the heart of a commonly used stadium roof called the hanging dish; the 202-foot-wide concave, concrete roof that tops New York City's Madison Square Garden is one such example. The arena, which is named for a bit of urban geography rather than geometry, is composed of a circular reinforced concrete wall that supports a radial spiderweb of steel cables embedded in concrete. The rigidity of the roof is derived from stiffening the cables by prestressing. The radiating cables are attached in the middle of the roof to a tension ring and at the perimeter to the circular wall. To prestress the cables, the builders first placed the prefabricated wedge-shaped concrete slabs that would make up the surface of the roof. Once they were in place, a ballast of bricks was balanced atop the slabs, stretching the cables. At this point the slabs and cables were "welded" with a cement mortar which, when it hardened, prevented the cables from returning to their original state. The result is a prestressed monolithic dish-like roof that is much stiffer than if it had been stabilized only by its own weight.

THE LARGEST TENTS IN THE WORLD

The array of enormous tension structures includes modern variations of the venerable tent. The roof which protected Roman spectators from the sun when they gathered in the Colosseum was a retractable canvas and rope awning, manned by soldiers, and spanning part of the arena's 512-foot width. The engineers who devised this basic shelter had no alternatives but canvas, rope and other natural materials. However, these are not ideal materials for large and permanent suspended roofs because they deteriorate in moisture and sunlight. Nor are they able to withstand high tension over great distances; the rule of thumb for material roofs is that the wider they are, the tauter the membrane must be stretched. At the same time, the risk of tearing increases with tension. Twentieth-Century builders also use the tent idea to cover vast spaces, replacing rope with steel cable and natural fibers with acrylics; these materials are preferable because of their greater tolerance of the elements and their much higher resistance to tension. High quality rope, for instance, has a tensile strength of 12,000 pounds per square inch, while high tensile steel's is 225,000 or more.

Among the best expressions of this updated tent design are the beautiful, sweeping "cable and membrane" structures, many of which were designed by the German architect Frei Otto, beginning in the 1950s. In most buildings, dead weight

The Dulles Airport terminal, completed in 1962, features a graceful sloping roof that garners rigidity from the union of steel cables and concrete. The parallel cables run from a row of higher inclined pylons to a row of lower pylons. The concrete roof was poured, in formwork, around the cables, resulting in a rigid roof that acts as a single, thin-shelled unit. The tensile strength of the cables prevents the long span of the roof from dangerous bending.

Like a giant tent, a huge steel and acrylic roof covers the
Munich Olympic Stadium. Completed in 1972 for the games
held in Germany, the roof is a vast tension structure made up
of a wide network of steel cables running in two directions.
Tens of thousands of acrylic plates attached to the cables constitute the transparent outer skin, which is 89,700 square yards
in area. Nine towers and their heavy cables provide the main
support for the roof from above, while 49 additional masts
help to hold it up from below.

is a major governing factor for stability. In cable and membrane structures this factor is reversed. All roof, these structures are formed of thin, light membranes that apply little dead load to their supporting masts. However, the membranes, which are in pure tension, also have very little inherent stiffness. To prevent them from being blown away by winds, the roofs must be firmly anchored to the ground and kept in tension by cables attached around the perimeter. As well, they must be of an ideal shape that will not deform under any load. Otto designed tents made from a variety of fabrics, but it was the large structures fashioned from acrylic panels and cables that brought him international recognition.

For the 1972 Olympics in Munich, Otto created a spectacular crescent-shaped, 18-acre spiderweb supported by masts and cables above the 78,000-seat main stadium. Towering above its periphery are nine hollow steel masts from which the spiderweb is suspended by cables. Tens of thousands of thin translucent plates of tinted plexiglass, each only 1/6th of an inch thick and up to nine feet square, are fitted into the web for shade and shelter against rain and sun. Each plate is bordered with a thin strip of metal and sealed with rubber buffers which assure that the roof does not leak. The result is a vast and expansive surface, contoured by the peaks and valleys of the quilted plastic membrane, and held in tension by heavy cables at its boundaries.

A tent membrane is at its strongest and stiffest when the tension is distributed evenly, with no slack or overstretched areas. Achieving such distribution is impossible in a large expanse unless it is divided up into smaller sections like those created by the Munich stadium's web. The size of a membrane in relation to its support structure is also critical. If it is too large, some parts will be insufficiently tensioned. But these structures are not constructed simply by draping a fabric or acrylic membrane over a set of masts; they are complex designs. As with a simple hiking tent, the pieces of the roof need to be carefully tailored and cut to produce what is known as a "minimum surface," a form that cannot be achieved with a single large piece. On a scale as large as the Munich Olympic Stadium, mathematical precision in calculating the shapes and sizes of the pieces that will make up the roof is essential.

The beauty of cable and membrane structures is that, when calculated and constructed correctly, these roofs have the potential to cover enormous areas. A recent example is another remarkable Frei Otto structure known as Mountain Tents for Pilgrims. Completed in 1981, it is a 23-acre tent that shelters the Muslim faithful at Mecca during their annual pilgrimage to Saudi Arabia. And yet this may be small compared to the possibilities opened up by current and future technologies. Air-supported, cable-reinforced bubbles of strong, long-lasting fabrics have the theoretical capacity to span 6,000 feet—more than enough to enclose an entire mile-square town. Within the bubble people could live in an artificial environment superior to the natural one; an oasis could be created in a desert or a pleasantly temperate settlement in Antarctica. Optical properties engineered into plastics could endow them with variable transparency so that a roof could darken in bright sunlight or become more translucent at lower light levels. Intuition and reason, inventiveness and technology, art and science are bound to make the inner spaces of the future amazing places, as thrilling and beautiful to the crowds gathering there as the landmark spans of the last 2,000 years.

SKYSCRAPERS

Throughout the world, skyscrapers have become the structural symbol of 20th-Century cities. Towering expressions of prestige and power, skyscrapers represent the coming together of sophisticated technologies that only began to mature in the late 19th Century, mainly in the United States. Before that time, other structures—notably the spires of Europe's Gothic cathedrals—had soared to great heights, but none had so combined height with utility. Instead of empty space, these new towers enclosed offices, apartments and hotel rooms for everyday use. By doing so, they allowed far greater density of use atop small but expensive lots in city centers.

Strictly speaking, a skyscraper is a tower that is many times higher than its base. The earliest skyscrapers were buildings 10 stories high; the modern definition begins at 20 stories. Their very height imposes structural demands that make skyscrapers different. Before the advent of the skyscraper, tall buildings relied on their walls, whether stone, brick, wood or concrete, to transfer the dead and live loads imposed on a structure to its foundation. Accordingly, for every story added, the walls had to be more massive at the bottom. Height, therefore, was limited by how thick the walls could be at the base; a building of only 18 stories would need lower floors several yards thick, an impractical configuration that would use up valuable ground floor space.

Height, without excessive width, was made possible by a combination of new construction methods and new materials. The structural weight was removed from the walls and placed instead on a lighter metal framework which channeled all the loads to a foundation below. The former weight-bearing walls became a mere skin to protect the interior from the elements. These slender frames, or skeletons, were the products of the burgeoning iron and steel industries; by the turn of the century strong and reliable forms of iron were widely available and increasingly economical to manufacture. But there had to be other advances to spur the rise

---

Skyscrapers, such as these looming above downtown Los Angeles, are the tallest inhabited structures in the world. The highest soar over 1,400 feet—more than a quarter of a mile. So tall are these buildings that the outside temperature may vary as much as 20 degrees from base to crown—and elevator riders' ears often pop from the change in pressure.

of modern skyscrapers. Safe, speedy elevators were needed to carry people to new levels inaccessible to the average stair climber. Some form of efficient central heating had to be devised; otherwise, much of the building's space on the upper floors would be taken up by chimney flues. A reliable water supply and good plumbing facilities that worked at the new heights were essential. And the danger of fire—with its potentially catastrophic consequences in such a concentration of people and property—demanded fire-retardant materials and new fire-fighting equipment. Some of these elements began to appear in the mid-1800s, but in isolation, and were not integrated into single buildings until the end of the century.

STRONGER, SAFER, HIGHER

Though Manhattan claims some early examples of the genre, Chicago is generally acknowledged as the birthplace of the skyscraper. Devastated by the great fire of 1871, the booming commercial city was ripe for a rebuilding based on safer materials; at first iron protected by terra cotta seemed to fit the bill. Gradually, from about 1890 on, the skeletons of the new buildings were held together by a cage of steel beams and columns acting as a single structural frame. Not only had steel become cheaper during the last decade of the 19th Century, but also steel girders and beams had become available in a variety of shapes and sizes. The most common was the I-section beam—in cross section it looks like a capital "I"—which was shaped from a single ingot. The connections between the beams and columns were made with steel connector plates and rivets, heated until they were red-hot and then hammered into pre-drilled holes.

The 15-story-high Reliance Building, built between 1890 and 1895, was typical of the emerging Chicago style and a harbinger of the great skyscrapers to come. The outside of the building clearly reveals the form of the steel skeleton within because its exterior is largely glass. This independent skin—a feature that later came to be called a curtain wall—demonstrated the design possibilities created by weight-bearing frames. By freeing architects from the structural limitations of supporting walls, curtain walls allowed them to focus on other concerns, from lighting and ventilation to the building's visual "statement."

Manhattan's Woolworth Building, completed in 1913, was a clear expression of a new esthetic made possible by these engineering breakthroughs. Retail wizard Frank W. Woolworth wanted a dramatic "Gothic building" for his company's headquarters; architect Cass Gilbert used ornamentation to transform the tall, slender structure into a kind of modern steeple, soon dubbed the Cathedral of Commerce. The exterior was tiled with hundreds of thousands of terra-cotta blocks, in nearly 500 sizes and shapes. Inside was a massive steel frame, the heaviest in any building to that time; one girder alone measured 23 feet long and 6 feet wide and weighed 68 tons, massive by early 20th-Century standards. From its foundation, the Woolworth Building rose 792 feet, making it the tallest building in the world. It held that distinction for 17 years until 1930 when another New York high-rise, the Chrysler Building, was completed. Only one year later the honor passed to a third structure that is still the epitome of the skyscraper and may be for decades to come: the 1,250-foot tall Empire State Building.

Designed in 1929, in the earliest days of the Great Depression, the Empire State became a symbol of economic rebirth as construction of the building forged ahead

ANATOMY OF A SKYSCRAPER

A typical skyscraper, such as the one at right, is built around a rigid steel frame and rooted in a solid foundation. The foundation consists of piles that are driven into the ground in clusters and capped with reinforced concrete; the main columns of the building rest on these caps, which distribute the weight to the piles. The strong, load-bearing frame, which can also be made of concrete, allows the building to be extremely high without the need for enormous lower levels to support its massive weight. Although the frame appears to be made of vertical columns and lighter horizontal beams, when bolted and welded together, it acts as a single, monolithic unit. The skyscraper is further stiffened against high winds and earth tremors by means of a reinforced central elevator core.

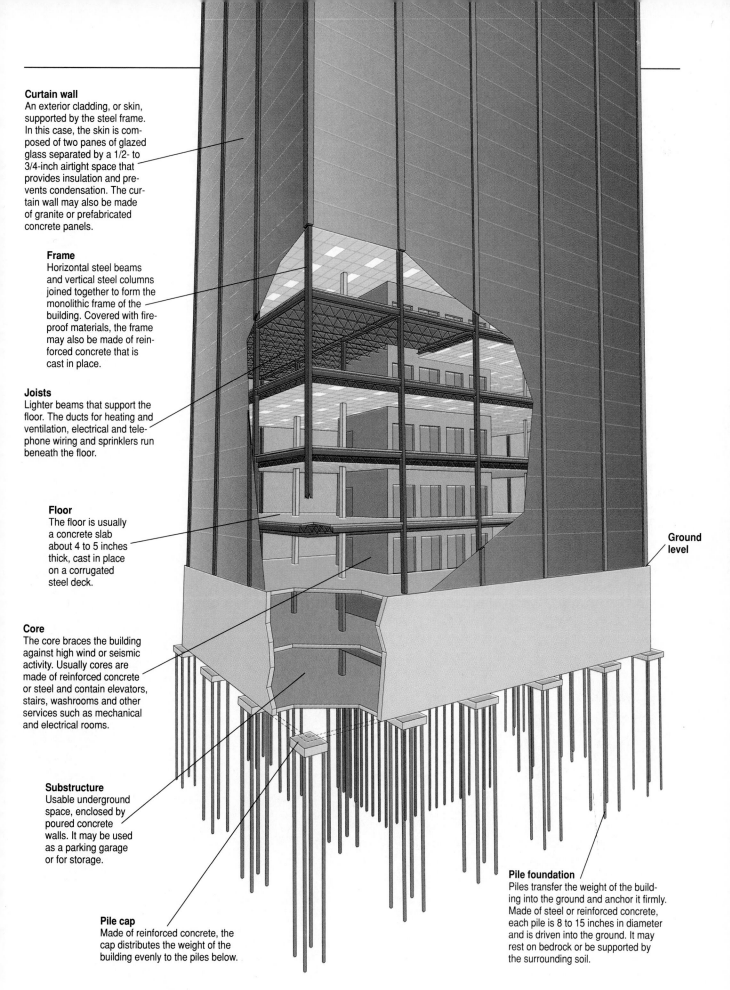

Curtain wall
An exterior cladding, or skin, supported by the steel frame. In this case, the skin is composed of two panes of glazed glass separated by a 1/2- to 3/4-inch airtight space that provides insulation and prevents condensation. The curtain wall may also be made of granite or prefabricated concrete panels.

Frame
Horizontal steel beams and vertical steel columns joined together to form the monolithic frame of the building. Covered with fireproof materials, the frame may also be made of reinforced concrete that is cast in place.

Joists
Lighter beams that support the floor. The ducts for heating and ventilation, electrical and telephone wiring and sprinklers run beneath the floor.

Floor
The floor is usually a concrete slab about 4 to 5 inches thick, cast in place on a corrugated steel deck.

Core
The core braces the building against high wind or seismic activity. Usually cores are made of reinforced concrete or steel and contain elevators, stairs, washrooms and other services such as mechanical and electrical rooms.

Substructure
Usable underground space, enclosed by poured concrete walls. It may be used as a parking garage or for storage.

Pile cap
Made of reinforced concrete, the cap distributes the weight of the building evenly to the piles below.

Ground level

Pile foundation
Piles transfer the weight of the building into the ground and anchor it firmly. Made of steel or reinforced concrete, each pile is 8 to 15 inches in diameter and is driven into the ground. It may rest on bedrock or be supported by the surrounding soil.

115

in the face of a collapsing world economy. Indeed, the great edifice was completed ahead of schedule and below budget because hard times had lowered labor and material costs. The 102-story, 60,000-ton steel frame was put together in less than 6 months, and the building was completed within 14 months at a cost of less than $25 million. At the peak of construction, there were 3,400 workers; by the end of the job 14 had died in accidents. Into the skyscraper went 120 miles of piping, 473 miles of electrical wire, more than a thousand miles of cable for the 73 elevators, and 6,500 windows. Its steel frame is covered with 200,000 cubic feet of cut limestone and granite as well as 730 tons of aluminum and stainless steel. Anchored in solid bedrock 55 feet below ground, the 365,000-ton Empire State Building has proved equal to all the forces that man and nature have thrown against it. Lightning strikes are a frequent occurrence, but the steel frame carries the current harmlessly to the ground.

Even before the Empire State Building went up, New York had begun to control the massing of skyscrapers with zoning laws. In lower Manhattan, the height and bulk of new buildings already threatened to turn the streets below into dark, cheerless canyons. These zoning laws dictated that the facade of high-rise structures be stepped back at various heights to provide light and ventilation. Architects of

A CENTURY OF SKYSCRAPERS

Skyscrapers evolved primarily in the American cities of Chicago and New York after 1850. By 1900, the fixed internal steel frame was the accepted design for tall buildings, and fast, safe electric elevators carried passengers to dizzying new heights. With each new landmark came improvements in foundations, frame construction, wind-bracing, fireproofing, elevators and other services such as plumbing, lighting and heating. The internal steel frame has persevered, although reinforced concrete is now commonly used. New designs will be for buildings of even greater height and stiffness, but that use less construction materials.

1913: Woolworth Building
New York. A 60-story tower, much taller than it is wide. Thick concrete piles were sunk 110 feet into the ground for a secure base. It boasted high-speed elevators controlled by a switchboard, and was a prominent example of the skyscraper as corporate monument.

1931: Empire State Building
New York. A 102-story building with a steel skeleton that was erected in record time. A good example of the set-back tower design necessitated by the 1916 zoning law.

1856: Haughwout Building
New York. A 5-story building with cast-iron construction and load-bearing walls; its steam-operated Otis elevator had a patented safety system.

1889: Tacoma Building
Chicago. A 13-story building with an iron skeleton frame, using load-bearing walls. The street fronts, however, are non-weight-bearing, or curtain walls, supported by the skeleton but independent from it. The frame is covered with terra-cotta for fireproofing.

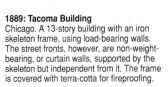

1890: Manhattan Building
Chicago. A 16-story building featuring a true iron skeleton, with no load-bearing walls. The frame was bolted together, but not considered a fixed, or stiff, frame. An early example of a structure with a wind-bracing system, using diagonal rods.

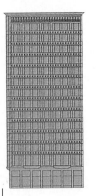

1895: Reliance Building
Chicago. A 15-story building with a fixed steel structural skeleton, riveted together. Lightweight terra-cotta external cladding made it a precursor of the glass curtain-wall buildings.

the day took the code literally, creating high-rise ziggurats that reinforced New York's reputation as a latter-day Babylon.

During the 1960s and 1970s steel-framed skyscrapers that far exceeded the height of the Empire State Building were built. As well, the quest to build cheaper and safer edifices led designers to turn more frequently to other materials, in particular, reinforced concrete. Much cheaper than steel, concrete is also far more resistant to fire because of its higher melting temperature and because the bulk of the material protects the steel reinforcement within. Like their steel counterparts, reinforced concrete frames also act like monolithic structures, enduring bending and compression as a single unit. Reinforced concrete was developed during the last part of the 19th Century and was the material used for the 16-story Ingalls Building completed in Cincinnati in 1903. But although it was commonly used in Europe beginning in the 1920s, reinforced concrete did not become common for the frames of high U.S. skyscrapers until the 1960s. Today, Chicago's Water Tower Place, completed in 1976, is the world's tallest reinforced concrete building at 859 feet. Now, buildings that use both steel and reinforced concrete are more commonplace. Regardless of their materials, all skyscrapers encounter similar loads and stresses, and must be designed to withstand them.

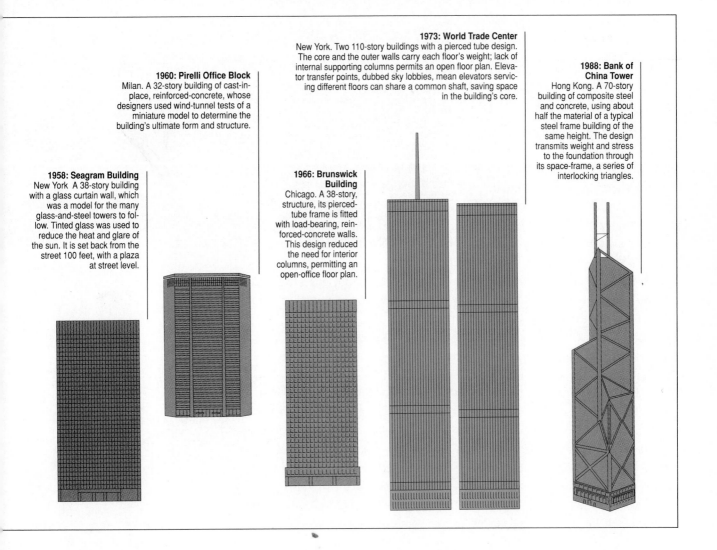

1973: World Trade Center
New York. Two 110-story buildings with a pierced tube design. The core and the outer walls carry each floor's weight; lack of internal supporting columns permits an open floor plan. Elevator transfer points, dubbed sky lobbies, mean elevators servicing different floors can share a common shaft, saving space in the building's core.

1960: Pirelli Office Block
Milan. A 32-story building of cast-in-place, reinforced-concrete, whose designers used wind-tunnel tests of a miniature model to determine the building's ultimate form and structure.

1988: Bank of China Tower
Hong Kong. A 70-story building of composite steel and concrete, using about half the material of a typical steel frame building of the same height. The design transmits weight and stress to the foundation through its space-frame, a series of interlocking triangles.

1958: Seagram Building
New York A 38-story building with a glass curtain wall, which was a model for the many glass-and-steel towers to follow. Tinted glass was used to reduce the heat and glare of the sun. It is set back from the street 100 feet, with a plaza at street level.

1966: Brunswick Building
Chicago. A 38-story, structure, its pierced-tube frame is fitted with load-bearing, reinforced-concrete walls. This design reduced the need for interior columns, permitting an open-office floor plan.

DESIGN AND TESTING

Structural engineers consider many factors when translating the client's needs and the architect's vision into concrete and steel. Naturally there are the rudiments of size, shape and weight, and the building's function. But beyond that, skyscrapers exist in an ever-changing environment. The loads created by the occupants of the building, busy urban traffic and weather are stresses that must be withstood on a daily basis. And great height means even greater susceptibility to gusts of winds or earth tremors from below.

Even non-gusting, insubstantial winds can be a source of trouble for a building made of heavy steel and concrete because a skyscraper can act like a very tall sail, catching the full force of breezes that become more powerful with increasing distance from the ground. Furthermore, the pressure the wind exerts increases with its velocity, so that relatively minor variations in wind speed can produce much larger pressure changes. In fact, for skyscrapers that are taller than 60 stories, the stresses produced by wind pressure cause a greater load than those exerted by the pull of gravity; in extremely tall buildings, wind may be responsible for as much as 75 percent of the total load exerted on the structure.

Wind produces a toppling effect, like the force of a puff of air on an upended pencil. Designers counter this force by sinking skyscrapers in strong foundations. But they have to cope with another effect as well—the tendency of a deeply rooted building to sway like a tree in a stiff breeze. A 1,000-foot-tall building in a wind of 100 miles per hour could sway by as much as four feet on its upper floors—enough to slosh the water in toilet bowls and to make motion sickness a new workplace hazard.

Engineers try to prevent such swaying by making buildings stiffer. The traditional method relies on the powerful and rigid steel frame, sometimes reinforced with diagonal braces, to shoulder the load. But today it is prohibitively expensive to build steel skyscrapers as massive as the Empire State Building which sways imperceptibly, even in strong winds. Modern builders try to do more with less, and have developed structural systems with greater strength-to-weight ratios. Some designs transfer wind loads from the outer walls to a heavily reinforced inner core that carries the force down to the foundation. Other systems rely on a stiff, external skeleton to deflect wind stress. The great X-shaped trusses in the outside walls of Chicago's John Hancock Center, for example, stiffen the structure with far less steel than would have been necessary in the internal frame; the cross-bracing also gives the building an unmistakable visual identity. Another Chicago skyscraper, the Sears Tower—at 1,454 feet, the world's tallest building—is in fact nine separate rectangular tubes of different heights. Bundled together and joined at their shared inner walls, the entire structure works as a single unit to absorb wind and seismic loads.

The designers of Manhattan's 915-foot Citicorp Center have dealt with wind-induced oscillation in a completely different fashion. A large room on an upper floor houses a remarkable device called a tuned mass damper, a 410-ton block of concrete on a movable platform that is connected to the sides of the building by arms that resemble huge automobile shock absorbers. When a computer senses the building beginning to sway excessively in a high wind, it activates pumps that force oil onto hydraulic footpads beneath the platform; the block floats on this frictionless layer of oil. Calculating the building's rate of oscillation, the computer sets the block in motion counter to the building's sway. The absorber arms control excessive movement by the block. When the wind dies, the computer drains the oil from the platform and lowers the concrete block to rest until it is needed again.

Wind-tunnel testing on models has become an essential element in the design of skyscrapers. The tunnel at left, 13 feet wide and almost 7 feet high, is capable of blasting winds of more than 60 miles per hour at scale models of skyscrapers. A magnesium carbonate powder, here blown from right to left through the tunnel, is used to visualize turbulence.

TESTING HYPOTHETICAL TOWERS

Like aircraft designers, engineers commonly test miniature models of proposed skyscrapers in the simulated world of the wind tunnel. Many of the systems that help buildings stand up to the constant pressure of the wind are also strong enough to withstand the brief but traumatic shock of an earthquake. Nevertheless, further provisions may be necessary to keep a skyscraper stable when the ground begins to move. In earthquake-prone areas such as Japan and California, building codes incorporate quake construction standards. Some buildings are equipped with instruments that measure structural stresses and displacements during earthquakes. More data comes from experiments with model buildings mounted on shaking tables—platforms as large as 20 feet square that are driven by motors to replicate the motion of an earthquake.

In recent years, engineers have been able to use this wind and seismic data to develop sophisticated computer modeling programs. With this software they can build a skyscraper on screen, specifying materials and dimensions, then subject it to the forces of a hurricane or an earthquake. Watching a video screen, they can monitor the building as the energy is transferred from floor to floor, through the frame, ceilings and walls. If it is improperly braced, a skyscraper can resonate like a tuning fork with the frequency of an earthquake and can multiply the ground forces as much as fivefold by the time they reach the top. By computer testing the building, engineers can reconcile such stresses with the strength of the building materials and the energy-dissipating characteristics of the design—without risking potentially disastrous trial and error.

When tests and models have shown that the basic elements of the design are sound, the detailed planning of the skyscraper can begin. This is a painstaking process, for all the systems in the building—electrical, mechanical, plumbing and so on—must be made to work together and, for the sake of esthetics, remain as unobtrusive as possible. The actual blueprint for a skyscraper can run to thousands of pages of detailed drawings—or gigabytes of computer memory. The structural plans alone must show the exact dimensions of each member in the steel frame, including the precise position of every hole for every bolt in every member.

Tour de force

As skyscrapers soar ever higher, transmission towers must rise above them to broadcast radio and television signals without encountering obstacles that might cause signal-bounce, or ghosting. The CN Tower, in Toronto, Canada, at 1,815 feet the world's highest freestanding structure, was built to eliminate this problem, but soon became much more than the means to a clearer TV picture. It is a symbol of Toronto and a tourist attraction, where 1.5 million visitors a year can see more than 100 miles in any direction from its 1,465-foot-high deck.

The reinforced-concrete structure, with its three wide hollow legs around a central tube, is structurally a three-legged stool. Unlike the more common cylindrical transmission towers, the CN Tower's distinctive design actually catches the wind, increasing wind stress. However, construction models were tested in a wind tunnel, and the tower was designed to withstand a 260-mile-per-hour wind. Wind sway is kept to an acceptable level by two spring-mounted 10-ton dough-nut counterweights that circle the antenna and act as dampers.

To anchor this giant securely in the ground, workers excavated 62,000 tons of material to reach a shale base 50 feet below the surface. They then built a 22-foot-thick foundation containing 9,200 cubic yards of concrete, reinforced with 500 tons of steel and 40 tons of tensioning cable. To erect the tower quickly—an average of 20 vertical feet a day—a slipform was used. A mold, or form, was built around vertical rods, and concrete poured into it. Once the concrete had set enough to support some weight, powerful hydraulic jacks pushed the form higher up the rods for another pour; this was repeated until the desired height was reached. The size of the slipform assembly was adjusted as it climbed to achieve the tapered tower shape. Owing to this great height and slim profile, the CN Tower tended to twist counter-clockwise due to the Earth's rotation, which could affect the fresh concrete. The slipform had to be monitored carefully and adjusted to compensate for this torsional oscillation. The finished tower deviates a mere 1.1 inches from plumb.

The Skypod, a seven-story, ring-shaped structure, was added at the 1,100-foot level on the tower to house lookouts, a revolving restaurant and broadcasting equipment; microwave dishes that receive communication signals are in its base. The Space Deck, containing the upper lookout, was installed, then a huge helicopter lofted into place the 39 sections of the 335-foot-high antenna, some of which weighed 8 tons.

Besides visitors, who have been ascending since it opened in 1976, this lofty testament to modern engineering also has lured daredevil climbers, parachutists and hang gliders.

Toronto's tallest skyscrapers are dwarfed by the 1,815-foot CN Tower, seen overlooking Lake Ontario. This height will keep its communications signals free of obstruction even if future buildings reach 1,000 feet.

FOUNDATIONS SHALLOW AND DEEP

Every cloud-touching skyscraper is also firmly anchored in the earth by way of a foundation that begins as nothing more than a giant hole in the ground. The foundation supports the dead weight of the building and its contents, as well as absorbing the forces exerted by earthquakes and winds. A solid rock foundation—the best base for skyscrapers—can support as much as 100 tons per square foot. In some locations, builders simply dig through overlying soil until they reach bedrock, dozens of feet below the surface; then they can begin to build back up. Broken bedrock or compacted gravel is also a fair weight-bearing base, withstanding as much as 10 tons per square foot. And, with the right kind of foundation, even the loosest of soils can be used to underpin a skyscraper.

The simplest foundation is called a mat, a wide pad of reinforced concrete, from three to eight feet thick, that spreads the weight of the building—snowshoe-like—over a broad area. The result is that no single bit of ground receives more weight than it can support.

A commonly used deep foundation employs steel or concrete piles hammered into the ground until they reach bedrock or a soil with sufficient bearing capacity. A variation of standard piles, called friction piles, need not reach solid rock. Driven into the ground in greater numbers than standard piles, they support their load by virtue of the friction created between the many piles and the ground. In both methods, the piles are driven in clusters, which are then capped with a concrete

STIFFENED AGAINST THE WIND

When skyscrapers were a mere 10 or 20 stories high, seismic activity—unpredictable earthquake movement—was the main concern for building engineers. Today, as the towering edifices grow ever higher, lighter, and more flexible, wind has become the most critical factor. For engineers, the forces of moving air and moving earth present the same structural problems. Wind twists and lashes a skyscraper from above. Earthquakes shake it from below. Unless these taller buildings have additional structural reinforcement to stiffen them, occupants could experience motion sickness or, as has happened in the past, weak structures could even collapse. At right are five innovative techniques that combat this dangerous swaying.

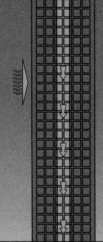

Central Core
The core, (above), found in all skyscrapers, is usually placed centrally in the structure. When the core is reinforced with vertical bracing, it acts as a backbone that stiffens the building against gusts of winds and earth tremors. It is the most common device in skyscrapers to combat dynamic loads.

footing upon which a single building column will rise. In some cases the piles may be driven several hundred feet deep.

However, even such clusters of piles may not be able to support the loads of widely spaced columns. Another type of deep foundation, called a caisson foundation, is better at handling this kind of load. A single caisson is a large-diameter steel casing that is filled with concrete to create a thick underground column. The casing may be inserted into a bored hole or actually driven into the ground, after which the silt, sand or gravel inside is excavated and replaced with concrete. Whether concrete or steel, each vertical member—or structural column—in a building rests atop a single concrete caisson.

FROM BELOW THE GROUND

The deep hole that marks the first stage of a skyscraper is a far more sophisticated construction than its disordered appearance from the street may suggest to an uninformed observer. Simply keeping the bottom dry enough to work, for example, often necessitates a complex water-removal system. Excavations may fill with water from rain or snow, an underground stream, or a high water table. In some cases, pumps may suffice to remove water from the hole, but if a water table is high, it must be lowered. The preferred method is to drill a system of dry wells around the site, all of which are deeper than the planned excavation. The water present in the site then flows into the nearby wells.

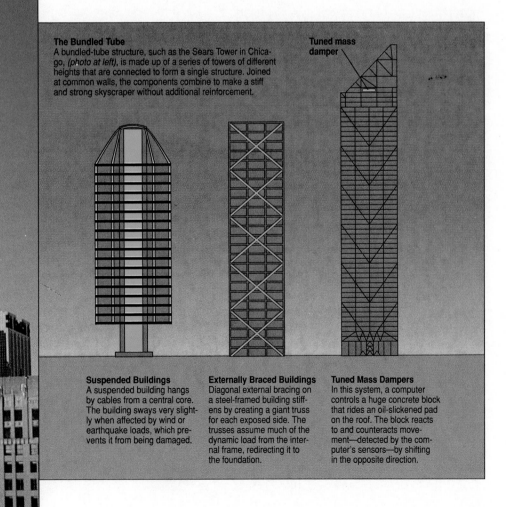

The Bundled Tube
A bundled-tube structure, such as the Sears Tower in Chicago, *(photo at left)*, is made up of a series of towers of different heights that are connected to form a single structure. Joined at common walls, the components combine to make a stiff and strong skyscraper without additional reinforcement.

Tuned mass damper

Suspended Buildings
A suspended building hangs by cables from a central core. The building sways very slightly when affected by wind or earthquake loads, which prevents it from being damaged.

Externally Braced Buildings
Diagonal external bracing on a steel-framed building stiffens by creating a giant truss for each exposed side. The trusses assume much of the dynamic load from the internal frame, redirecting it to the foundation.

Tuned Mass Dampers
In this system, a computer controls a huge concrete block that rides an oil-slickened pad on the roof. The block reacts to and counteracts movement—detected by the computer's sensors—by shifting in the opposite direction.

Tiebacks
Prevent the retaining wall from collapse; the tensioned cables are anchored to bedrock or solid ground and attached to horizontal retaining beams that are welded to the vertical steel columns of the retaining wall.

Retaining wall
Built around the perimeter of the excavation site to keep the surrounding ground from collapsing; typically made of steel columns driven into the ground every 6 feet to support horizontal sections of wood.

Tower crane
Erected at the perimeter of the site or inside the core of the building; used to lift heavy construction materials and equipment; the height of the crane is increased by adding sections to it.

Cement mixer
Delivers fresh concrete to the site and keeps it mixed by rotating it until needed; the concrete may be poured into a concrete bucket, as shown, or pumped to the appropriate location.

Columns
Made of concrete and reinforced with vertical steel bars, the columns support the weight of the floor above and transfer it to their footings.

Formwork
Made of wood and built to the specifications of a column or other structural member, the formwork is used to support concrete until it has set.

Column footings
Massive blocks of reinforced concrete that rest on bedrock or solid ground; the weight of the building is channeled down the columns, through the footings and into the ground below.

FLOOR BY FLOOR

A construction site affords curious pedestrians a lesson in how a skyscraper actually goes up. As the plywood hoarding around the site is raised, machines, materials and manpower begin to arrive. Though it may look chaotic, the building process is systematic, dictated by blueprint specifications and work schedules, and orchestrated tightly by a project manager.

The raising of the columns and floors of a reinforced concrete skyscraper, like the one illustrated below, begins after the foundation and substructure are completed. First, temporary wooden formworks, made of plywood sheets and planks, are built for the columns. Reinforcement bars are placed in the formwork and concrete is poured around them. Once the concrete

sets, the formwork is removed and recycled. Next, a similar process creates formwork for the floor, which is laid over a supporting scaffold and covered with a mesh of steel reinforcing bars. Concrete is poured over the mesh and spread evenly across it, usually to a depth of four to eight inches. And foot by foot, floor by floor the building reaches for the sky.

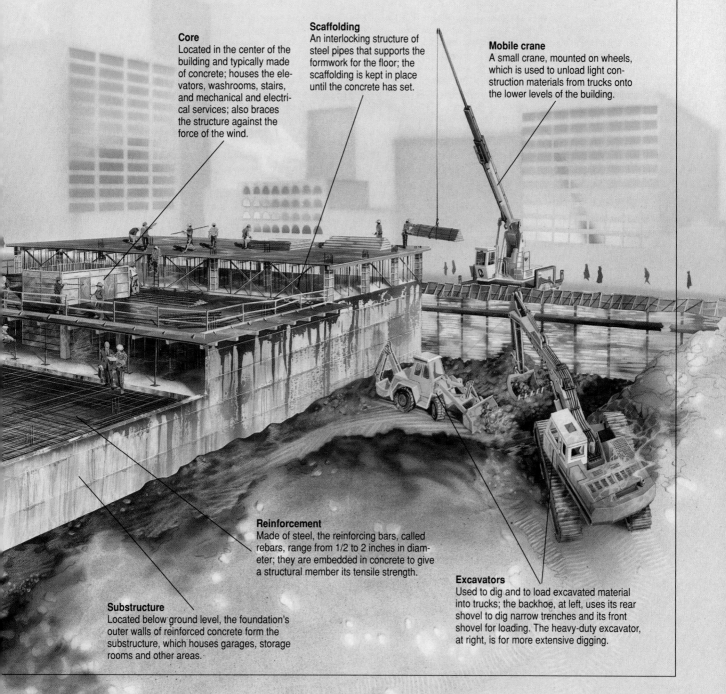

Core
Located in the center of the building and typically made of concrete; houses the elevators, washrooms, stairs, and mechanical and electrical services; also braces the structure against the force of the wind.

Scaffolding
An interlocking structure of steel pipes that supports the formwork for the floor; the scaffolding is kept in place until the concrete has set.

Mobile crane
A small crane, mounted on wheels, which is used to unload light construction materials from trucks onto the lower levels of the building.

Reinforcement
Made of steel, the reinforcing bars, called rebars, range from 1/2 to 2 inches in diameter; they are embedded in concrete to give a structural member its tensile strength.

Excavators
Used to dig and to load excavated material into trucks; the backhoe, at left, uses its rear shovel to dig narrow trenches and its front shovel for loading. The heavy-duty excavator, at right, is for more extensive digging.

Substructure
Located below ground level, the foundation's outer walls of reinforced concrete form the substructure, which houses garages, storage rooms and other areas.

The activity at a construction site early in a project contributes to the apparent chaos. First, several kinds of digging equipment attack the soil: Broad-bladed dozers loosen and push it into piles; loaders use their wide scoops to lift it into huge hauler trucks; power shovels and backhoes dig deep with narrow scoops, dumping their loads onto haulers or a nearby spoil bank for later hauling or as backfill around the finished foundation. If rock needs to be removed to reach the desired depth, blasting teams drill holes, tamp explosives, cover the site with heavy rubber mats, then set off the rock-crushing detonations. Digging equipment removes the rubble, and blasting teams return to break up the next layer of rock.

As the excavation goes down, temporary retaining walls around the perimeter of the site go up to prevent cave-ins. One common type of wall uses sheet piles—corrugated steel panels, driven into the floor below the level of the excavation. Another kind of wall uses regularly-spaced steel I-beams driven into the ground; a solid face of horizontal timbers fitted between the supports holds back the earth. To brace retaining walls against the pressure of earth and water, builders commonly drive hollow steel rods, called tie-backs, through the walls and diagonally into the earth behind. Concrete is injected into the tube and forms an anchor when it spills out and hardens at the other end.

When the digging is done, the building begins. If the structure is to rest on piles or caissons, builders drive them through the bottom of the pit to the desired depth. Otherwise, the first concrete floor, which will become the lowest level of the substructure, is poured. This first level will distribute the column loads from above to the piles or caissons below, or throughout the solid rock foundation. From this concrete pad, the skeleton of the new skyscraper will begin to rise.

WORKING WITH STEEL

Despite the increased use of concrete in high-rise construction, the typical skyscraper in the making is still a soaring frame of structural members made from steel. Within this steel frame, hundreds of feet above the street, work crews methodically assemble the thousands of pieces that will transform it into a monolithic tower. The steel members they work with—usually made of specially manufactured structural steel—take many shapes. Typically a steel member is "I"- or "H"-shaped in profile. The cross piece of the H or the I is called the web; the flat sections on either side of the web are known as flanges. The flanges of steel beams or columns can range in width from 4 to 36 inches and they can be up to 5 inches thick. The heaviest and thickest members are the square H-shaped vertical columns which are typically one to two stories in height. The I-beams used for horizontal members are manufactured at whatever lengths are required.

The steel for a skyscraper is fabricated to order in mills; standard shapes are rolled and cut to length, then ironworkers weld or rivet brackets for the joints and drill what are still called rivet holes, even though bolts have long since displaced rivets in on-site construction. The consignment is usually delivered to an interim storage area, each piece with a painted identification number to tell construction workers its exact placement in the structure.

The steel arrives at the construction site only a few hours before it is needed, both to avoid exposure to the elements and because the huge members take up a great deal of room in an already crowded working area. Ironworkers loop cables

Termite Towers

Home owners may know them as destroyers of wooden buildings. But in nature, termites are among the most remarkable builders and architects known. Many Australian and African termite species build veritable miniature skyscrapers—some as high as 20 feet. Inside are complex warrens of access tunnels, chambers and ventilation systems; outside is a hard external shell which protects the earthen edifice.

Towers such as these house millions of working and breeding insects. They are constructed by the worker termites from soil mixed with saliva; this natural concrete creates the rigid external shell. A network of channels inside the towers circulates air; tiny ducts in the walls can be opened or closed by the insects to exchange stale air for fresh. The towers, built to last, are maintained and repaired continually; although the lives of the insects may span only a year, their towers may stand for a century.

The plain exterior of this Australian termite tower belies the complex within. The hard external walls protect a Queen's chamber, fungus gardens and millions of termites. The structure's wide base, which extends several inches below ground, assures stability.

under individual columns or beams—or a group of them—hook the cables to a crane and then lift the steel off the truck. For lower stories the crane may extend from a motorized vehicle on the ground; for higher jobs a tower crane is commonly built up through the center, as the building becomes taller, where the elevator shaft will eventually go, or sometimes along the side of the building. When it is no longer needed, the crane is dismantled in sections; if it was in the center of the building it is replaced with an elevator.

The tower crane places the steel temporarily on the highest working level. There, raising gangs—the crews responsible for erecting the steel frame—often use smaller cranes rigged at their working level to move the individual columns or beams into place in the frame. Some members of the raising gang attach lifting cables to the steel, while others help guide it as the smaller crane lifts it through the framework. The process is controlled by hand signals or, for higher buildings, radio contact with the crane operator. But the most visible ironworkers in the team are the two connectors, who frequently balance on a beam or column high over-head while guiding another member into its correct position. The horizontal steel beam, with its flanges facing up and down, will be attached at a right angle to the vertical face of a column. The actual connection between the two main members is made with the aid of at least two small steel brackets manufactured at 90° and called angles. Like the columns and beams, the angles have had rivet holes drilled through them at the steel mill. The workers use the pointed tails of tools called spud wrenches to line up rivet holes of the columns, angles and beams, then use the tool's wrench head to secure them temporarily by means of a single bolt. If the steel members do not fit flush and at the right angles, as occasionally happens, the members are separated and work stops while one of the crew fires up an acetylene torch and shaves away excess metal. Once the beam is secured, one connector edges out onto it to release the lifting cable so that the crane can bring up another piece of steel. The pace of the progress varies from building to building, but the completion of one floor every three days is not uncommon.

A crew of "bolters-up" follows the raising gang, putting bolts with washers into all the rivet holes and threading nuts loosely onto them. But before they can be tightened, a surveyor must check to be sure the steel is precisely aligned—both vertically and horizontally. Using precision instruments, the surveyor signals for adjustment of any steel that is out of place. The ironworkers stretch wire cables equipped with turnbuckles between columns, then tighten or loosen the turnbuckles until the steel is pulled into line. Only then are the bolts secured, with a 30-pound air-powered impact wrench. The wrench is set to apply a specified torque, or turning force, to the nut, and to stop turning when the designated stress is reached. Structural bolts are classified into two types: carbon-steel or high-strength. Carbon-steel bolts are cheaper but not as strong as high-strength bolts, and are used for connections where lower strength is sufficient—usually in the interior of the building. High-strength bolts, which are heat-treating during the manufacturing process, are used for the main frame connections, which must withstand tremendous shear stress. After installation, an inspector checks each bolt to make sure that it has its proper washers and the correct torque.

Next, workers equipped with electric arc welders move in, further securing the structure by welding the members together along thin lines at the connections. Usually individual columns are two stories tall; they are erected from the working floor, then joined by beams at the top. The steel columns are mated vertically, end to end, by means of flat plates that are bolted against the flanges of the column. The higher a building rises, the lighter the columns can be become since they will be supporting less load. After the columns and beams of the main frame are trued, bolted and welded, horizontal joists, the lighter supporting beams that will hold the floor, are bolted in place. Corrugated steel sheets, called Q-decking, are laid atop the joists to make a new working floor where the steel for the next level will be stored. Then, while one crew erects the next set of columns from this working floor, another assembles the framework of the beams for the floor below. When all the connections are made, the entire frame of columns, beams, bolts and welds works as a single, rigid unit.

After the ironworkers have passed, teams of workers move onto the floors. Over the Q-decking they lay miles of conduit for electrical wiring, plumbing, telephone

Working high above a construction site, a steel worker bolts together one of the steel beams and columns that comprise the rigid internal frame of a modern skyscraper. The steel members are positioned with cranes and loosely connected with bolts by workers using hand tools. After a series of adjustments have assured that each member is perfectly straight and level, the frame is permanently bolted together with a power tool called an impact wrench.

cable, even computer networks. Next, workers install wire mesh reinforcement across the whole floor surface, and a raised frame around its edge, before pouring four to five inches of concrete. The concrete, freshly mixed, is brought to the construction site in the rotating drums of huge concrete delivery trucks. Sometimes the trucks dump their loads into the hopper of an enormous concrete pump, which can deliver concrete through extension pipes to hard-to-reach places in the foundation and on the lower floors. High levels, however, are out of reach of even the most powerful pump; concrete must be hoisted there in oversize dump buckets. Rubber-booted workers work quickly to push the soupy concrete into the mesh, level it according to the directions of a surveyor, and smooth it out with both hand and power trowels.

The bolting of the frame and the pouring of the floors allows work on the exterior walls to begin. In some buildings, the curtain wall begins as a grid of metal strips fastened to the framework; windows and wall panels are then installed between the members of the grid. In other structures, huge sections of the

A modern skyscraper, such as Chicago's 74-story-high Water Tower Place, completed in 1976, functions as a small, self-contained city, also known as a multi-use complex. Three glass elevators service a seven-story shopping mall on the lower floors—including shops and theaters—which is wrapped around a central atrium. Floors above the mall also contains offices, an entire hotel and condominiums. The different facilities are served by separate elevator systems and lobbies. The entire structure has more than 610,000 square feet of floor space.

wall—sometimes including windows—are bolted directly to the frame. Once all the panels are installed, the wall is made weatherproof by filling the outside joints with a caulking compound. This flexible material does double duty by also serving as a buffer between panels, taking up the slack as the separate sections expand and contract with changes in temperature and the effects of wind and weather. Within the curtain wall, crews install insulation and other barriers that will prevent fire from spreading between floors.

FINISHING THE JOB

When curtain walls and windows are complete, the skyscraper looks finished from the outside, but an enormous amount of work remains within. The walls that break up the vast floors into corridors and rooms are normally constructed of gypsum drywall board fixed to lightweight steel studs. Heating and cooling plants are installed, and the metal ducts that will carry the temperature-controlled air throughout the building are suspended beneath the concrete floors. Below the ducts, hanging from wires attached to the floor above, are metal grids for drop ceilings; the grid is eventually filled in with tiles and fluorescent lighting fixtures that create the effect of a solid ceiling.

In the elevator shafts installers assemble the machinery that will allow elevators to carry passengers at speeds up to 20 miles per hour while stopping no more than a fraction of an inch off the mark at each floor. Each car rides within a framework of steel guide rails, suspended from steel cables. In the event of a cable failure, safety clamps on the car engage the guide rails, preventing a fall. In the tallest buildings, the elevator system takes on the complexity of a small railroad. Express elevators carry passengers to sky lobbies high in the building, where they transfer to local elevators that stop at each floor. Each tower of the World Trade Center in New York has two such lobbies, at the 44th and 78th floors; 99 elevators in numerous shafts serve the different levels. This helps reduce the amount of valuable space devoted to elevators; another means to the same end is the double-deck elevator installed in some buildings.

Almost a small town in its own right, a skyscraper requires abundant and reliable electricity to keep running. Heavy-duty circuitry and wiring carry the current to the elevator motors, the air conditioning plant and other mechanical uses, as well as to distribution boards on each floor. There the current is apportioned to the various circuits for lighting and appliances; these circuits are made of lighter wire running through conduits already cast into the concrete floors. The floors also hold many of the local supply and drain pipes for fresh and waste water. Drainage systems usually run straight down the building's service core. Water supply, however, requires a more complicated arrangement in a skyscraper. The pressure of the water main can usually push water only about eight stories up; above that, pumping is necessary. To avoid having to run pumps constantly, many buildings are designed to have water distributed by gravity flow from rooftop tanks to which the water is initially pumped in stages 10 to 20 stories at a time; the tanks are topped off by pumping whenever their contents fall below a predetermined level. In other buildings, the fresh-water system is fed by smaller pump-filled tanks placed at eight-story intervals. This allows a consistent and strong water pressure to be maintained throughout the edifice.

The 1,209-foot-high Bank of China Tower, in crowded downtown Hong Kong, may point the way to the future for many skyscrapers. The building is based on a space-frame design, taking advantage of massive, three-dimensional trussing rather than a traditional internal steel frame. The entire weight of the building is borne by four corner columns. Moreover, the materials used are a combination of structural steel and concrete composites. The result is a lighter, less expensive structure with the same strength and rigidity as traditional skyscrapers of similar size.

STATE OF THE ART

Contemporary skyscrapers, for all their impressive height, may only hint at the towers of the future. The engineering challenges of taller skyscrapers give rise to the need for strong, lightweight, economical construction methods. The 1,209-foot-high Bank of China in Hong Kong is Asia's tallest building and is an example of one new direction for steel structures. The skyscraper's skeleton is based on a triangular trussed grid—known as a space frame—rather than the traditional rectangular frame. It required 50 percent less steel than a conventional frame, significantly cutting the building's total cost.

Concrete, which is the preferred construction material in many of today's skyscrapers, will continue to have a place in the buildings of the future. It is cheaper than steel, and its mass and rigidity produce stiff structures. Research continues into combining concrete or steel with the kinds of plastics and composite materials commonly used in the automobile and aerospace industries, where weight considerations are important factors.

Because there are no indications that skyscrapers will cease to appear on cityscapes around the world, there is concern in some quarters about the consequences of their continued proliferation. With their immense weight concentrated on a small area, skyscrapers can cause excessive strain on already overloaded infrastructures beneath the streets. The massing of these towering edifices can also block out light and surrounding scenery from urban cores. New York's World Trade Center sometimes casts a shadow that is two miles long. Skyscrapers also create concrete canyons that generate their own wind conditions and patterns, creating problems for pedestrians as well as for the structures themselves. Alternatives to building high must be found.

At the same time the problem of creating business space in highly populated downtowns may, in many cases, override these concerns. Architects continue to design—and engineers continue to build—tall skyscrapers. But the growth in height and volume of the tallest has been restrained. Cities such as San Francisco, Seattle and Boston already have instituted tough controls that limit the height or location of skyscrapers and, for the moment, the 1974 Sears Tower at 110 stories, or 1,454 feet, remains the world's tallest building. Engineers claim that the technology and know-how exists to build edifices 50 or more stories higher and that there soon may be a sudden surge at the high end of the scale in both the United States and in Asia. Since the legendary architect Frank Lloyd Wright first discussed the idea of a "mile-high" skyscraper, there have been an assortment of designs unveiled for incredibly high edifices; such structures, however, still seem fanciful.

But the real challenge lies not so much in the construction of the skyscrapers, but in making them serviceable; bigger buildings mean more people, sometimes tens of thousands more, who require increased transportation facilities, utilities and services such as police and daycare. Higher towers require elevators that can ferry people quickly and facilities that ensure rapid access and escape in case of emergency. Indeed, the amount and type of planning and coordination involved in constructing a modern skyscraper would seem to be more in keeping with planning an entire city than a single building. Nevertheless, the allure and romance of trying to create the tallest buildings in the world seem to be as provocative and challenging as ever.

Material Gains

It is interesting that the design and construction of even the most modern structures frequently rely on old ways. Many of these principles are as trustworthy and enduring as a well-engineered concrete arch.

This is the case for the new materials that will underpin the future of structures. Tried and true, concrete and steel will continue to be the dominant construction materials for years to come. But improved versions are appearing; a new heat-treated, high-strength steel is 20 to 30 percent stronger than ordinary steel and will allow the construction of lighter structures without sacrificing size. And extensive research is being carried out into composites—the combination and integration of two or more components into a single material, stronger and more durable than its components. Usually one of the elements is in the form of fibers woven throughout the material. Yet this, too, has roots in the past. In nature, for example, trees garner strength from the combination of long cellulose fibers in a matrix of lignin, or natural polymer. And reinforced concrete, which has allowed the construction of countless long, high and wide structures, is essentially a composite. When reinforced by thin steel bars, concrete works as a single material.

But concrete is vulnerable to the chlorides found in road salts, and even the steel embedded inside can corrode. Recently, researchers have experimented with infusing concrete with long fibers of polypropelene or glass. The laboratory results are a concrete that is corrosion resistant, 100 times more flexible and 4 times stronger than standard concrete—attributes that would serve structures well in earthquake-prone areas. Also holding great promise are materials rein-forced with carbon fibers. Carbon is the non-metallic, natural element that is at the heart of steel. Composites composed of an expoxy-resin base reinforced with carbon fibers has proved to be remarkably resistant to corrosion and fatigue. These are strengths that would enhance the durability of cable-stayed bridge cables, which suffer particularly from problems such as these.

Innovative research in composites is aimed at producing "smart" materials with implants embedded in concrete or steel that will reveal degradation before it becomes serious. Eventually, materials may diagnose their own faults and alert engineers to the need for repair or replacement.

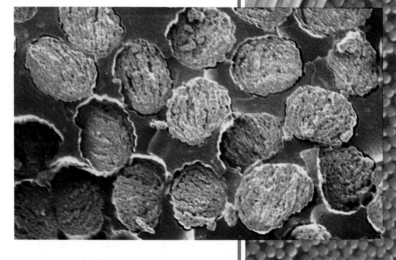

Future materials are seen under an optical microscope. A glass, carbon and aramid composite, magnified 300 times, reveals the layered construction of its fiber and polymer matrix. Inset is a 3,000x-magnification of carbon fibers in a polymer matrix.

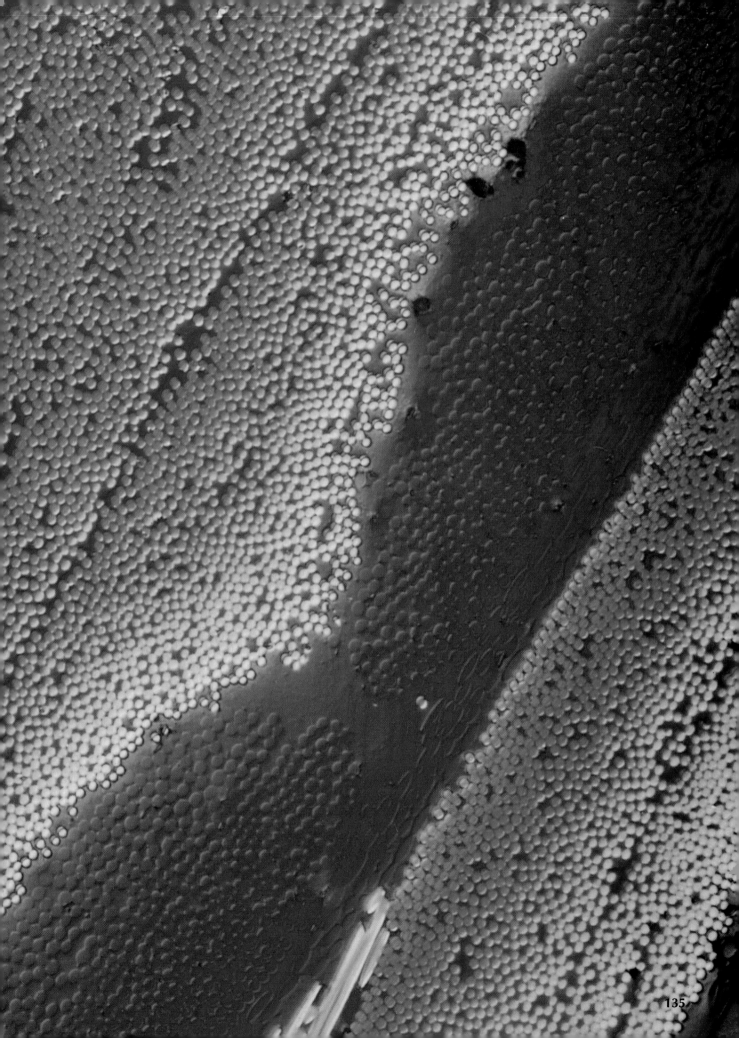

New Tools of the Trade

By and large, the methods of the building trade changed very little until the 19th and 20th Centuries brought new materials and tools to the work site. While building a large structure is still mostly a manual operation, two new tools are changing the process. Computers already have utterly changed the design and testing stage; and robotics are well on their way to altering the very nature of construction.

Computers have increased the speed of design and testing manyfold. Massive data bases provide architects and engineers with access to vast amounts of information and storage—all at a keystroke. Now, new architectural software and advanced graphics allow people to "walk" through a structure long before the first shovel of earth is turned. Viewers can zoom into on-screen images to examine subtle nuances or alter details. Researchers currently are working on an even more useful software tool known as Virtual Construction. Here designers will be able to render a structure, or any element of it, alter it as needed, and test it under loads or under the simulated forces of nature. The response will be almost instantaneous.

At the worksite, humans may eventually be replaced altogether for some jobs. Construction robots have been a long-elusive goal, but one that is coming incrementally closer. Functioning prototypes for automated machines that can apply plaster and spray on fireproofing are already in existence. Other

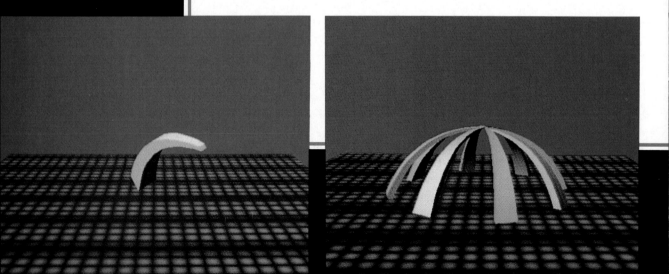

robots do more discrete jobs, such as installing ceiling panels, studs for drywalls or blocks for masonry wall construction. However, the rugged and dirty nature of many work sites presents more of a problem for fully automated robots; building a machine that can negotiate the unpredictable terrain of a construction site is like trying to teach a baby to walk. Architects and engineers will have to plan for their use during the design phase. But the day of the robot is fast approaching. Remote-controlled machines already are in limited use. Internally programmed robots for specific jobs are in an advanced development stage. And artificially intelligent robots that will make their own considered decisions are a dream approaching reality.

With an experimental computer program known as Virtual Construction, designers will build, alter and test structures on the screen. The first two images show how ribs are made into a dome. The images below show how the dome can be tested under a hypothetical load from above. The same process can be applied to other forces, such as wind loads, and takes only minutes.

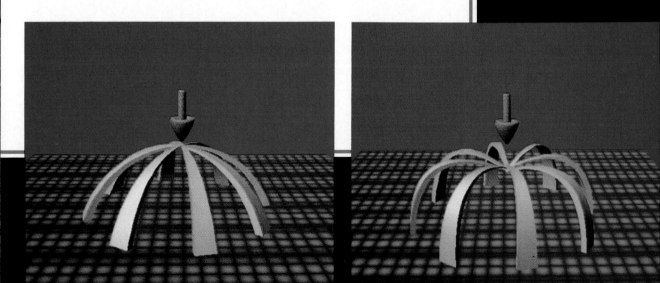

Possible Dreams

Yesterday's dreams of amazing structures are often today's realities. A tunnel beneath the English Channel had long been discussed, but for many years it seemed an absurd idea. Today, the first pilot tunnels have met beneath the channel, and trains and cars will be speeding between France and Britain by 1993. Now serious consideration is being given to a bridge over the Straits of Gibralter—an entirely feasible project, spanning a mere nine miles.

But many such dreams have gone unrealized. In the 1950s, architect Frank Lloyd Wright proposed a mile-high skyscraper. Other designers worked on similar structures but, while such edifices are technically possible, they remain impractical.

Some dreams have been partially realized. In 1925, architects unveiled *The Titan City, a Pictorial Prophesy of New York, 1926-2026* in New York City. The exhibition was a vision of a city with stepped-back skyscrapers, connected by transit systems, elevated walkways and bridges. Zoning laws allowed for plenty of sunlight and room to move around. It was an optimistic dream of a new urban order. But though the technologies to build such structures developed apace and skyscrapers went up, the urban planning necessary for such utopias was less successful.

Today there are new dreams. On the drawing boards in Japan is a conical artificial city more than half a mile high. Houses, commercial and recreation facilities, offices, school, theatres, parks and waterways all would be enclosed in one cloud-skimming structure called Sky City. The designers envisage 14 internal "space plateaus" connected to each other by three-story-high elevators. Spiral monorails will carry people up into the structure from a huge parking zone and unmanned buses will zip them around the concave plateaus. Each plateau will have its own electrical power station; wind power and solar energy will be used as well. Composite carbon materials that are up to 50 percent lighter than reinforced concrete will be used for construction of the six-million-ton structure. And to complete the picture, the builders of this extraordinary structure of the future will be robots.

Sky City 1000 is one vision of future structures. This 3,250-foot tower would accommodate 35,000 residents and 100,000 office workers. Sky City would have twice the surface area of New York's Central Park, but would take up only about one-thirtieth of the land space.

Index

Numerals in *italics* indicate an illustration of the subject mentioned.

PICTURE CREDITS

Multiple credits on a page are read left to right, top to bottom, divided by semicolons.

Cover: Photograph by Graeme Outerbridge

6 Dave Lawrence/The Stock Market. 7 Wilhelm Schmidt/Masterfile; Anders Kristoffersen/MCS of Scandinavia A/S. 14,15 Liam Hubbell/Woodfin Camp & Associates. 26,67 Lou Jones/The Image Bank. 29 Courtesy Computers & Structures Inc. 32 Hans Pfletschinge/Peter Arnold Inc. 35 Gerry Cranham/Photo Researchers Inc. 36,37 Graeme Outerbridge. 40,41 Janeart Ltd./The Image Bank. 42 Courtesy Honshu-Shikoku Bridge Authority. 44,45 Dave Lawrence/The Stock Market. 47 Courtesy Honshu-Shikoku Bridge Authority. 48,49 Courtesy Honshu-Shikoku Bridge Authority. 50,51 Hydro Quebec. 56,57 M. Austerman/Animals Animals. 59 Lowell Georgia/Photo Researchers Inc. 60,61 Courtesy National Rivers Authority, Thames Region. 65 David M. Stanley/Photo Researchers Inc. 68,69 PRISMA/Sonderegger. 74,75 Brian R. Wolff. 77 Courtesy Atlas-Copco. 78,79 Gary G. Gibson/Photo Researchers Inc. 82 Anders Kristoffersen/MCS of Scandinavia A/S. 84,85 Nicolas Fievez; QA Photos Ltd. 87 Janeart Ltd./The Image Bank. 88,89 Brian Masck/All-Sport USA. 93 Daily Telegraph/Masterfile. 94,95 Ian Steer/SkyDome. 96,97 Mike James/Photo Researchers Inc.; Courtesy Sydney Opera House. 99 Robert Chartier. 100 Eric Bouvet/Gamma-Liasion. 103 Wilhelm Schmidt/Masterfile. 104,105 Deborah MacNeill, courtesy B.C. Place. 106,107 Ned Gillette; Bob Howe (2). 109 Arthur Murphy. 110,111 Alain Choisnet/The Image Bank. 112,113 Pete Saloutos/Masterfile. 118 Philippe Railly/Photo Researchers Inc. 120,121 Bill Brooks/Masterfile. 122,123 Steve Elmore/The Stock Market. 127 Bill Bachman/Photo Researchers Inc. 129 Mike Dobel/Masterfile. 130 John Bryson/The Image Bank. 132 P & G Bowater/The Image Bank. 134,135 Thomas R. Clark and Michele E. Pavis/Lehigh University; Thanasis Triantafillou/Massachusetts Institute of Technology. 136,137 Prof. Alex Pentland/Massachusetts Institute of Technology.

ILLUSTRATION CREDITS

80,81 Robert Monté. 124,125 Shawn Potvin/L. Madore, concept by Yves Pelland. 138,139 Courtesy Takenaka Corporation

ACKNOWLEDGMENTS

The editors wish to thank the following:

Fredrick Allen, American Heritage of Invention and Technology, New York, NY; Atlas Copco Construction and Mining N.A., Montreal, Que.; Tony Baez, American Society of Civil Engineers, New York, NY; John E. Bower, ATLSS Center, Lehigh University, Bethlehem, PA; Jonathan Cherneff, Center for Construction Research and Education, MIT, Cambridge, MA; CN Tower Ltd., Toronto, Ont.; John Corrigan, Port Authority of New York and New Jersey, New York, NY; James Cox, The North Face, Berkeley, CA; Oscar Dascal, Montreal, Que.; Paul Dykes, National Rivers Authority, Thames Region, Reading, England; John G. Everett, Center for Construction Research and Education, MIT, Cambridge, MA; Virginia Fairweather, Civil Engineering, New York, NY; Dal Freeman, Arrow Dynamics, Inc., Clearfield, UT; Ned Gillette, Ketchum, ID; Atsushi Goto, Honshu-Shikoku Bridge Authority, Tokyo, Japan; John W. Hager, Structural Engineer, Savannah Army Corps of Engineers, Savannah, GA; Brad Hankinson, Nexus Technology Inc., Tokyo, Japan; Charles Helliwell, Center for Construction Research and Education, MIT, Cambridge, MA; Chris Hendrickson, Carnegie-Mellon University, Pittsburgh, PA; Bob Howe, Oakland, CA; Irene Javor, British Columbia Hydro and Power Authority, Vancouver, B.C.; Norma King, Consulate-General of Great Britain, Montreal, Que.; Ray Lablonde, British Columbia Place Stadium, Vancouver, B.C.; Michel Lee, Brookfield Development Corp., Montreal, Que.; Lovat Tunnel Equipment Inc., Etobicoke, Ont.; David A. Marks, A.I.A., Loebl, Schlossman and Hackl, Inc., Chicago, IL; Massachusetts Port Authority, Boston, MA; Mitchell Merowitz, Toronto Transit Commission, Toronto, Ont.; Metropolitan Atlanta Rapid Transit Authority (MARTA), Atlanta, GA; Said Mirza, Dept. of Civil Engineering, McGill University, Montreal, Que.; Ed Moloney, Vollmer Associates, New York, NY; Jane Morley, Department of History and Sociology of Science, University of Pennsylvania, Philadelphia, PA; Nashville Division, U.S. Army Corps of Engineers, Nashville, TN; Andrew Olmsted, S-Technology, Cambridge, MA; Claude Pasquin, Pasquin St-Jean et Associés Inc., Montreal, Que.; Alex Pentland, The Media Lab, MIT, Cambridge, MA; Rhaetian Railway, Chur, Switzerland; Leslie E. Robertson Associates, New York, NY; Bill Sampson, The Indiana Convention Center & Hoosier Dome, Indianapolis, IN; Charles Seim, P.E., T.Y. Lin International, San Francisco, CA; John Simons, Marathon Realty, Montreal, Que.; Russ Sinha, U.S. Bureau of Land Reclamation, Denver, CO; David Stringer, Montreal, Que.; Curtis D. Summers, Inc., Cincinnati, OH; David A. Thomas, ATLSS Center, Lehigh University, Bethlehem, PA; Peter Trepanier, Canadian Centre for Architecture, Montreal, Que.; Thanasis C. Triantafillou, Dept. of Civil Engineering, MIT, Cambridge, MA; Dr. Ruth D. Turner, Mollusc Dept., Museum of Comparative Zoology, Harvard University, Cambridge, MA; Masato Ujigawa, Takenaka Corp, Tokyo, Japan; Diana Wall, South Street Seaport Museum, New York, NY.

The Great American Scream Machine is a registered trademark of Six Flags Corporation.

The following persons also assisted in the preparation of this book:
Naomi Fukuyama, Stanley D. Harrison, Jenny Meltzer, Brian Parsons, Shirley Sylvain

This book was designed on Apple Macintosh® computers, using QuarkXPress® in conjunction with CopyFlow/CopyBridge™ and a Linotronic® 300R for page layout and composition; StrataVision 3d.®, Adobe Illustrator 88® and Adobe Photoshop® were used as illustration programs.